SpringerBriefs in Applied Sciences and Technology

PoliMI SpringerBriefs

More information about this series at http://www.springer.com/series/11159
http://www.polimi.it

Davide Spallazzo · Ilaria Mariani

Location-Based Mobile Games

Design Perspectives

POLITECNICO
MILANO 1863

Davide Spallazzo
Department of Design
Politecnico di Milano
Milan
Italy

Ilaria Mariani
Department of Design
Politecnico di Milano
Milan
Italy

ISSN 2191-530X   ISSN 2191-5318   (electronic)
SpringerBriefs in Applied Sciences and Technology
ISSN 2282-2577   ISSN 2282-2585   (electronic)
PoliMI SpringerBriefs
ISBN 978-3-319-75255-6   ISBN 978-3-319-75256-3   (eBook)
https://doi.org/10.1007/978-3-319-75256-3

Library of Congress Control Number: 2018931443

Printed on acid-free paper

This Springer imprint is published by the registered company Springer
International Publishing AG part of Springer Nature
The registered company address is: Gewerbestrasse 11, 6330 Cham, Switzerland

Preface

This book frames Location-Based Mobile Games from a design perspective, investigating the peculiar traits that make them compelling contemporary practices and challenging fields of investigation.

Bridging the real and the digital into novel and unexpected hybrid worlds, these games are emerging as powerful scenarios for exploratory processes and versatile means of communication, in addition to be recognised means of entertainment.

Relying on an interdisciplinary theoretical background and empirical studies, this book digs into Location-Based Mobile Games intertwining theoretical assumptions and descriptions of their translation into practice. The authors look at these games from different perspectives, exploring how they can impact on the way we look at our surroundings, their influence on our social dimension, their ability to translate a wide range of information into game experience and the negotiations they activate by intertwining two realities. Each issue is addressed from a twofold perspective: that of designers who craft the games and that of players who interpret the choices of designers and take part to the game experience. In so doing, the book covers the relationship between processes of designing and playing, investigating games that communicate through meaningful interactions, shares perspectives as forms of narratives and integrates physicality and surroundings in the play activity.

The reasoning advanced throughout the chapters is intended for researchers, designers and entrepreneurs in the field, since the book provides a novel perspective on Locations-Based Mobile Games, aims at affecting designers' awareness about issues often neglected and suggests interpretations and practices of use that can impact on the way commercial games are designed. Addressing this specific field of research, it provides a thoughtful perspective that adds experience-based knowledge to the current literature on Locations-Based Mobile Games. We propose a novel approach regarding the role of LBMGs in fostering informal situated learning and in favouring social engagement. Furthermore, we reflect on these games as persuasive media, with a significant narrative dimension, as well as on the function that physical objects can play within game experiences in hybrid spaces.

From a design perspective, stressing in particular the relationship between theory and practice, we invite game designers to reflect on the implications of its design choices in terms of game experience and transferring of meanings.

In conclusion, considering the entrepreneurs in the field, the book may be a source of suggestions, interpretations and practices of use that can be inspiration for proposing front-end solutions. Exploring game design features and techniques, tools and methods, it may support game production and development. In so doing, it can impact on the way commercial games are both designed and employed.

Milan, Italy   Davide Spallazzo
                                                                 Ilaria Mariani

Acknowledgements

This book is the result of various years of research and exploration, conducted largely benefitting from the invaluable support and contribution, expertise and exceptional talent of others.

As ideas and insights made their appearance, we had the opportunity to discuss them with brilliant colleagues and accomplished professionals and to test them with enthusiastic novices, as well as with skilled designers.

We found ourselves surrounded by many colleagues and friends who sustained our need to investigate, test and understand. Hence, we gratefully acknowledge those who challenged our insights, inspiring us and shaping our work. In this regard, our university has provided a fertile and supportive environment. We are deeply thankful to the School of Design and the Department of Design of the Politecnico di Milano, which allowed and encouraged our researches.

A special thank goes to the BS students of the course "Augmented Reality and Mobile Experience", A.Y. 2013/2014, 2014/2015 and 2015/2016. They put themselves to test by participating in the design and assessment activities, critically analysing their own projects and giving us constant, precious feedback.

This book owes a significant debt to those who collaborated with us and those who co-authored our past and ongoing works.

For their outstanding support, we offer special thanks to the significant people in our lives, friends and family.

Contents

Chapter 1
Introduction

Abstract The chapter introduces the book explaining its purposes and significance, framing it within the current literature related to Location-Based Mobile Games. It further clarifies the methodology of the study on the ground of this work and summarizes the content of each chapter.

Keywords Location-Based Mobile Games · Game Design · Persuasive Games
Methodology

Location-Based Mobile Games are coming of age. Ultimately, it has been a few years ago that digital games moved from the arcades to our houses with console and PC games. Afterwards, they became portable, then invaded the World Wide Web, immediately after digital games entered our mobile phones, and now they are expanding in the everyday space.

Location-Based Mobile Games (LBMGs henceforth) provide contextual play activities, since players are geolocated and transported into a hybrid world that blurs the real one with the digital. Since the introduction of 3G networks and the embedding of GPS in feature phones and smartphones, designers explored ways to expand digital games in the contingent world, giving birth to a hybrid space that intertwines the digital and the real realms.

It is not a coincidence that one of the first things that spring to mind speaking of games is the huge popularity and revenue of video and computer games, and their prominent role in the current cultural landscape. For instance, concerning digital entertainment, 63% of US households have at least one person who plays video games regularly that means 3 h or more per week (ESA 2016).[1] According to the global sales, the revenues coming from this market already overtook the one of Hollywood industry. This happens in a context where the gaming industry revenues

---

[1]Data from the 2016 *Essential Facts About the Computer and Video Game Industry*. Released by the Entertainment Software Association (ESA) in April 2016, this annual study is the most in-depth and targeted survey of its kind, and surveys a pull of more than 4000 American households about their game play habits and attitudes. Full document available at: http://essentialfacts.theesa.com/Essential-Facts-2016.pdf. Accessed 29 December 2016.

© The Author(s), under exclusive licence to Springer International Publishing AG,   1
part of Springer Nature 2018
D. Spallazzo and I. Mariani, *Location-Based Mobile Games*,
PoliMI SpringerBriefs, https://doi.org/10.1007/978-3-319-75256-3_1

are projected to grow by 35% over the next five years to $70 billion annually. Several times, it has been said that games became the media paradigm of the twenty-first century (Rifkin 2000), often surpassing television and film in popularity (Flanagan and Nissenbaum 2014). In parallel to the AAA worldwide known productions, a consistent proliferation of indie games (independent games) has progressively established, creatively exploring a rich spectrum of topics, experiences and perspectives in (video) games. Once again, this trend and its attitude towards issues of social matter mined the old and by now outdated negative stereotype of games as a pointless waste of time (Gee 2003; McGonigal 2011). If on the one side it affected the conventional definition of what is a game, on the other it raised reflection on the mainstream perception of its diverse and salient role as a contemporary medium. Games are today an engaging way to share other perspectives through a first-person experience: they are a way to put issues in the player's hands with potential interesting outcomes in terms of understanding and reflection.

Additionally, games' dynamics are increasingly applied to non-gaming contexts, entering the field of the so-called *gamification* (Deterding et al. 2011).

1.1 Purpose and Significance

The increased presence of gamified systems shows that processes of ludification are becoming more and more mainstream, attesting the relevance and integration of games in our daily life, with evident implications in the fields of marketing, communication, business, health and education (Deterding et al. 2011; Walz and Deterding 2015; Werbach and Hunter 2012; Zichermann and Cunningham 2011). It is safe to say that the pastime will keep us entertained, and in parallel that the ongoing trend requires investigation.

Considering the significance of the topic and the variety of perspectives it includes, we look at the existing practices and state of the art, tapping into the relationship between these games and the surrounding context, as well as its social implications. Moreover, we draw special attention to the function of narrative that some games tend to underestimate or to deal with without taking full advantage of its power and potentiality in terms of engagement and involvement. This consideration, shaped as a critique, comes from our perspective on LBMGs, an approach that interprets them as communicative tools and systems for conveying meanings, being it in the fields of education or learning, entertainment or public interest. This book is not devoted to the production and dissemination of fundamental knowledge about LBMGs in general; on the contrary, it relates to a particular epistemology that comes from practice-based research and results into a better understanding of practical uses and limits, aiming at bridging research and practice.

From a theoretical point of view, our contribution relies on concepts from the disciplines of game studies, communication design and interaction design, digging into transversal forms and functions that can be useful in educational and working settings. Grounding on the design field, a discipline overtly based on the practice of crafting, constructing and testing, we keep a holistic approach to the topic, considering LBMGs in their entirety, from the ideation to the deployment, passing through iterations of test and implementation. Therefore, moving back and forth between theory and practice, this book is structured as a mixture of theoretical assumptions and descriptions of their translation into practice. Recalling the concept of Constructive Design Research advanced by Koskinen and colleagues (2011), we informed the essay around the concept of *through-design* field research, translating in words an approach that credits as relevant going after the design artefacts in the real context of use. Methodologically, the book resumes and discusses some research topics we explored in our respective doctorates—in which games have been enquired as communication means—and reports on the empirical advances conducted over a three-year analysis on three B.Sc. courses.

In so doing, the book reaches out to a multidisciplinary approach that covers the relationship between processes of designing and learning, investigating games that communicate through meaningful interactions, share perspectives as forms of narratives, and integrate physicality and surroundings in the play activity.

Each chapter proposes a timely contribution to the existent literature that appears today fragmented, and mostly characterized by an approach that targets case studies analysis, or advances theoretical speculations about LBMGs impact. In consequence, and acknowledging the current needs emerged during our applied research, we adopted an approach that swings between theory and practice, where the first informs and nurtures the second and vice versa.

Summarizing what triggered this contribution, we can say that beyond the explorations we conducted in our respective research investigations, it was indeed the didactic experimentation that brought to bear some relevant points that are in need of exploration, as well as specific heuristic models we designed as complementary to standard design practices. Reasoning, perspectives and models are what we intend to share through this book.

1.2 Background of This Book

Whereas our investigations have revealed that both designing and playing games are powerful sources of understanding and even enculturation (Mariani and Spallazzo 2016; Mariani and Ackermann 2016), and that various meanings can be embedded into games (Spallazzo and Mariani 2017), we structured the contents of this book around three nodes: theory, design and social dimension.

The book counts upon our empirical research in the Politecnico di Milano, School of Design educational context, and benefits of a variety of experimentations we conducted over the years. As said, empirical experimentation grounded on existent theory that further impacted on our knowledge, bringing hands-on awareness and understanding.

To conduct focused examinations and collect a comprehensive base of validated, tested and documented knowledge, we followed the approach framed as "design as research" (Laurel 2003; Cross 2006; Schön 1983). The contribution of this book grounds on the existent literature and is nurtured by our design activity, as source of practice-based reflective insights. It is the result of a through-design research (Findeli 1998; Jonas 2007; Koskinen et al. 2011) that counts on a series of 44 LBMGs conceived, designed and prototyped at Politecnico di Milano, School of Design. Although the role of theoretical contributions is central and unquestionable, a parallel and doubtless function is covered by the practice of doing, in this case the above-mentioned act of conceiving, designing and questioning LBMGs, and the process of gaining awareness and understanding that may follow. We look at the design practice beyond these games as means to construct knowledge (Koskinen et al. 2011), as well as means for learning, entertainment and communication. Ignoring these games' (potential) ability to engage and also transfer knowledge means missing important opportunities to promote amusement, engagement and growth.

We are interested in exposing the diverse perspectives that make LBMGs interesting, fascinating and versatile communicative means of communication. Even if most of the reasoning and results we obtained can be extended to LBMGs in general, it is necessary to stress that we applied our research on a particular and challenging typology of games: persuasive LBMGs with social aims. We investigate games that put actual values and meaning-making at play that invite, suggest or provoke players to deal with controversial contents, critical issues, ethical matters, combining the reflection about playing games with making games into one connected and thoughtful reasoning.

1.3 Methodology

Acknowledging the current state of the art in the field, we focus on the results of the empirical research we conducted analysing the processes of both designing and playtesting 44 persuasive LBMGs. They are the outcomes of three assessments in the B.Sc. course "Augmented Reality and Mobile Experience" (a.y. 2013/14, 2014/ 15, and 2015/16). In this period, we involved a total amount of 180 students, asking them to look at their surroundings, get fascinated (for better or worse) by actual societal problems or taboos and evolve them into concepts to be translated into LBMGs. Our aim was to investigate how these games function as engaging communication systems able to entertain players and convey information in the meanwhile. Ranging from examining *design* as a process of enquiry, to considering *playing* as a way to gain knowledge, we observed and studied our sample of persuasive LBMGs conducting ethnographic analysis and interpretive research. Acknowledging that every research method has limits, biases and weaknesses (Denzin and Lincoln 2011; Creswell 2008), we used a mixed methods approach, collecting multiple forms of data and applying a triangulation of different methods. During the iterative cycles of design that spanned over three consecutive months for each academic year, we ran interpretative ethnography and participant observation

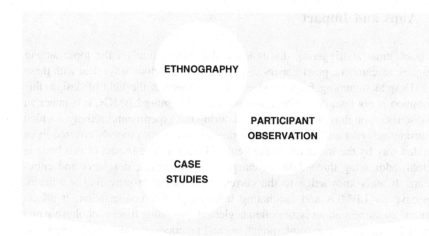

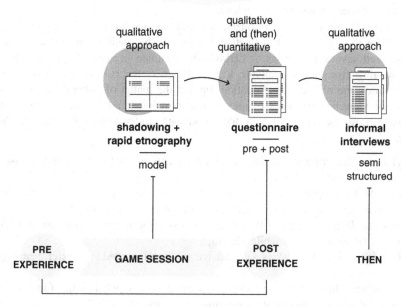

Fig. 1.1 Strategies and tools used to observe games and collect results

(Stake 1995; DeWalt and DeWalt 2001). In parallel, we conducted rapid ethnographies, shadowing, questionnaires and informal interviews with students (Fig. 1.1).

Especially conducting moderate participation (DeWalt and DeWalt 2001), we understood on the one hand designers' needs in terms of knowledge and tools, on the other how players received games and made sense out of them. In doing so, we grasped an important amount of insights regarding the relationship between initial expectations and effective players' perceptions.

1.4 Aims and Impact

This book aims at triggering discussion and confrontation on the topic among colleagues, researchers, practitioners and those who in various ways deal with these issues. Despite stemming from a didactical experience in the field of design, this contribution is not intended to be a handbook for designing LBMGs. It is rather an open discussion on those topics—emerged during the experimental activity—which we consider relevant and pertinent, since they are often not properly covered in an integrated way by the literature on the field of LBMGs. The impact of this book is threefold, addressing three different categories: researchers, designers and entrepreneurs. It adds knowledge to the current field literature providing a different perspective on LBMGs and facilitating interdisciplinary conversation; it affects designers' awareness about issues often neglected, providing timely scholarship and useful reflections; it suggests interpretations and practices of use that can impact on the way commercial games are designed.

As both theorists and practitioners, we came across the fact that games must account for the specific and distinctive features of the medium. In particular, we encountered several properties of LBMGs that resulted in need of specific reflection, ranging from tools available to the designer, the role of narrative-based architectures, the player–object relationship and the matter of cultural biases. As a result, we identified six areas of interest that turn into guiding questions for our observation, on the one hand, and discussion with students and colleagues on the other.

1. What are the critical foundations and intersections in the existent interdisciplinary literature?
2. How do we empower designers/players to recognize, grasp and theorize the results of their design/play process in terms of informal learning?
3. How can designers create LBMGs able to foster social engagement and how do players build social ties while playing?
4. How does a LBMG work as a persuasive medium intended to transfer meanings or trigger awareness?
5. Considering the communicative aim of these games, what is the function of narrative and its structure in conveying information?
6. How do LBMGs relate to the context/playground and what kind of interactions do in-game elements activate?

The questions listed above have been answered through dedicated chapters.

Chapter 2 *LBMGs in a Nutshell* critically frames LBMGs in the context of mobile-based experiences, addressing their development, their technological implications as well as the practices they enable and their impact on how we look at urban and public spaces.

Chapter 3 *Informal Learning Between Design and Play* addresses the practice of designing and the activity of playing as matters of learning out of formal education activity. Theoretically, we frame LBMGs learning outcomes in the fields of mobile learning, situated learning and game-based learning, focusing then on the process of meaning-making we observed in our didactic activity.

Chapter 4 *The Social Dimension of Located Play* addresses LBMGs as collaborative activities. Relying on theoretical assumptions and inferences drawn from practice-based research, we interpret LBMGs as triggers of social engagement within the group of players and with non-players.

Chapter 5 *LBMG as Persuasive Medium* looks at games as means of communication and as contexts of meanings, namely spaces where designers can codify, represent and perform meanings that players are asked to interpret and decode, making sense of the game and its layers of sense.

Chapter 6 *Stories, Metaphors and Disclosures: A Narrative Perspective Between Interaction and Agency* explores the role of narrative in LBMGs. It discusses how games take the shape of storyteller's point of view that translates information into game experience, taking advantage of metaphors, keyings and stealth approaches as ways to transfer embed messages and meanings without sacrificing players' enjoyment. Avoiding the unconscious activation of psychological barriers, a conscientious use of narrative is indeed functional to communicative impact.

Chapter 7 *Beyond the Digital: Reflecting on Objects and Contexts* addresses the role of physicality in LBMGs exploring how game objects may translate the fictional world into the real one, acting as boundary objects that intertwine the two worlds, rather than simply overlapping them, and activate negotiation of meaning. Furthermore, the chapter analyses how the physical context may influence the game experience and how it may be included by designers in LBMGs.

Although all the chapters reflect the authors' common and shared view on LBMGs, Chaps. 1 and 7 have been edited by the authors together, Chaps. 2, 3 and 4 have been written by Davide Spallazzo, and Ilaria Mariani edited Chaps. 5, 6 and 8.

References

Creswell JW (2008) Research design: qualitative, quantitative, and mixed methods approaches. SAGE Publications Inc, Thousand Oaks, CA

Cross N (2006) Designerly ways of knowing. Springer, London, UK

Denzin NK, Lincoln YS (eds) (2011) The SAGE handbook of qualitative research, 4th edn. SAGE Publications Inc, Thousand Oaks, CA

Deterding S, Dixon D, Khaled R, Nacke L (2011) From game design elements to gamefulness: defining "gamification". In: Proceedings of the 15th International Academic MindTrek Conference: envisioning future media environments. ACM, New York, NY, USA, pp 9–15

DeWalt KM, DeWalt BR (2001) Participant observation: a guide for fieldworkers. AltaMira Press, Walnut Creek, CA

ESA Essential facts about the computer and video game industry. In: The Entertainment software association. http://www.theesa.com/about-esa/essential-facts-computer-video-game-industry/. Accessed 29 Dec 2016

Findeli A (1998) A quest for credibility: doctoral education and research in design at the University of Montreal. In: Doctoral Education in Design

Flanagan M, Nissenbaum H (2014) Values at play in digital games. The MIT Press, Boston, MA
Gee JP (2003) What video games have to teach us about learning and literacy. Palgrave Macmillan, New York, NY
Jonas W (2007) Design research and its meaning to the methodological development of the discipline. In: Michel R (ed) Design research now. Birkhäuser Basel, pp 187–206
Koskinen I, Zimmerman J, Binder T et al (2011) Design research through practice: from the lab, field, and showroom. Morgan Kaufmann, Burlington, MA
Laurel B (2003) Design research: methods and perspectives, 1st edn. MIT Press, Cambridge, MA
Mariani I, Ackermann J (2016) Fun by design: the game design activity and its iterative process as (playful) learning practices. Conjunctions Transdisciplinary J Cult Participation 3:1–20
Mariani I, Spallazzo D (2016) Empowering games. Meaning making by designing and playing location based mobile games. ID&A Interac Des Archit 28:12–33
McGonigal J (2011) Reality is broken: why games make us better and how they can change the world. Penguin Press, New York, NY
Rifkin J (2000) The age of access: the new culture of hypercapitalism, where all of life is a paid-for experience. TarcherPerigee, New York, NY
Schön DA (1983) The reflective practitioner: how professionals think in action. Temple Smith, London, UK
Spallazzo D, Mariani I (2017) LBMGs and boundary objects. Negotiations of meaning between real and unreal. In: Proceeding of the 6th STS Italia conference
Stake RE (1995) The art of case study research. Sage Publications, Thousand Oaks, CA
Walz SP, Deterding S (2015) The gameful world: approaches, issues, applications. MIT Press, Cambridge, MA
Werbach K, Hunter D (2012) For the win: how game thinking can revolutionize your business. Wharton Digital Press, Philadelphia, PA
Zichermann G, Cunningham C (2011) Gamification by design: implementing game mechanics in web and mobile apps, 1st edn. O'Reilly Media, Sebastopol, CA

Chapter 2
LBMG in a Nutshell

Abstract The chapter critically frames Location-Based Mobile Games in the context of mobile-based experiences providing different definitions and a brief history of their development over time. It addresses the characteristics of LBMGs and their technological implications, discussing them as location-based services. It further analyses the impact of these games on the way we look at urban and public spaces, intending LBMGs as pervasive games, thus able to expand the player's experience in time, space and sociality.

Keywords Location-Based Mobile Games · LBMGs History · Geolocation
Pervasive Games

Location-Based Mobile Games (LBMGs), as the name clearly suggests, are games which exploit mobile device location awareness to provide players with a contextual game experience. Location is therefore the key point of this kind of games that modify the gameplay according to players' current location, mixing a digital experience provided through a mobile device with a physical one, performed in the real world.

LBMGs are frequently mentioned in the literature with other names such as *mixed-reality games* (Flintham et al. 2003; Montola 2011) or *hybrid-reality games* (de Souza e Silva and Delacruz 2006), focusing on their characteristic of being played between the digital and the physical worlds, with digital contents overlapping reality. The origin of these locutions can be traced back to the *virtuality continuum* of Milgram and Kishino (1994) who proposed a taxonomy of mixed-reality visual displays ranging from reality to virtuality, passing through augmented reality and augmented virtuality, being the first a digitally augmented representation of reality and the second, on the contrary, a virtual environment augmented by real objects.

Sometimes, this typology of games is also addressed as *augmented-reality games* (Squire et al. 2007; Jacob et al. 2012) underlying the process of digital augmentation of the real world, or referring specifically to games that superimpose AR graphics on reality, through the smartphone camera.

D. Spallazzo and I. Mariani, *Location-Based Mobile Games*,
PoliMI SpringerBriefs, https://doi.org/10.1007/978-3-319-75256-3_2

Although locutions such as *mixed-reality*, *hybrid-reality* and even *augmented-reality* are clearly appealing and vivid, in the book we chose location based to anchor them back to their very origin—the exploitation of location awareness—and consequently to keep the definition as inclusive as possible. Our discourse is indeed independent of the graphical quality and characteristics of the hybridization between reality and virtuality through the smartphone camera, and it focuses rather on the peculiarity of these game, that of impacting on the digital world through the physical one and vice versa.

2.1 Brief History of LBMGs

The origin of LBMGs dates back to the first years of the millennium, with the release of GPS-enhanced mobile phones. The games *Botfighters* and *Mogi* can be considered the precursors of LBMGs, since they are the first mobile games to rely on players' location for the gameplay. Released in 2001 by It's Alive as an urban Massively Multiplayer Online Role-Playing Game—MMORPG—*Botfighters* allowed players to locate other players in the city and engage in a battle to destroy them. In *Mogi*, released by Newt Games in 2003, players, provided with a GPS-enabled phone, had to look for and collect virtual objects located and disseminated in the real space and visualized on the mobile screen. Despite their simple gameplays and basic graphic design, *Mogi* and *Botfighters* introduced a novel way of conceiving video games, by bridging the physical and the digital worlds and compelling players to move in the real world to impact on the digital one.

Noticeably, the origin of the two above-mentioned games can be traced back to *Geocaching* (O'Hara 2008). Spread since 2000 thanks to the improved accuracy of GPS systems (Kelley 2006), the simple game consists in placing/hiding a *geocache*—a waterproof box containing a logbook and tradable items—in a specific place and diffuses over the Web its geographical coordinates in order to let other *geocachers* find it, register their passage and trade items.

In the same years, the potential of geolocated mobile phones was also employed to create playful urban experiences, event games, that temporarily transformed the city into a playground. A well-known example is *PacManhattan*, created in 2004 by Frank Lantz and his students of NYU's Interactive Telecommunications Program. The regular grid of the original game board was transferred into the streets and avenues of Manhattan around Washington Square Park so that players dressed as Pac-Man or ghosts could run to catch virtual dots or Pac-Mac, respectively. Each player was guided via cell phone by human controllers, who could check their position in real time through custom software on a PC.

The same years saw the development of several prototypes in research projects: *Can you see me now?*, born by the collaboration between the Mixed Reality Laboratory of the University of Nottingham and Blast theory (Benford et al. 2006), is one the most cited examples. Designed as a chasing game, it featured virtual runners on a map at the PC and real runners on the street: the aim of the street runner was to catch the PC runners, who, in turn, had to escape from them, while listening to the breath and voice of the runners (Montola 2011).

Other experiences developed in the first decade of the new millennium focused on the educative power of LBMGs transferring their characteristics to serious games. In 2002, the MIT Teacher Education Program directed by Eric Klopfer firstly experienced serious LBMGs on PDAs—Personal Digital Assistants: in the game *Environmental detectives* (Klopfer et al. 2002) groups of students participated in a real-time simulation based on a local watershed. Few years later, the same research group, in collaboration with the University of Wisconsin–Madison, created *Mystery at the Museum*, a LBMG aimed at providing visitors of the Boston Museum of Science with a novel and engaging learning experience (Klopfer et al. 2005).

In 2005, the first pilot of *Frequency 1550* (Raessens 2007; Akkerman et al. 2009; Huizenga et al. 2009) was released as the result of a common effort by Waag Society, KPN, IVLOS (University of Utrecht) and ILO (University of Amsterdam). Aimed at teaching young pupils the history of Medieval Amsterdam, the game involved players in location-based quests and mini-games to conquer city zones while moving across the downtown.

Two years later, *REXplorer* (Ballagas and Borchers 2007; Ballagas et al. 2007, 2008) embraced the same challenge, employing a LBMG to teach the story of a place: by picking quests from ghosts encountered across the city, players were led to discover Regensburg and learn about its past.

The fast diffusion of modern smartphones since the release of iPhone in 2007 (comScore 2011) and the following release of Android OS in 2008 favoured the development of LBMGs which quickly flourished for both the app markets. GPS-enabled mobile phones with wide touchscreens rapidly entered the houses, and developers could easily make their game available on the two markets. As a consequence, LBMGs saw a turn towards commercial products, and several games were released. A non-inclusive list encompasses *Gbanga* by Millform AG, *Gowalla* by Gowalla inc., *GPS Mission* by Orbster, *CityHunters* by the homonymous company, *Codecrackers* by La mosca, *Monopoly City Streets* by Tribal DDB and Hasbro, *Dokobots* by Dokogeo, *foursqWAR* by Project Zebra, *DarkCity 2029* by Jihad games, *Merchant* by Oberon Interactive BV.

Although most of the titles above mentioned are still alive and available today, commercial LBMGs in most cases were played by a niche of people and did not reach the wider public. An exception is the game *Ingress* (Majorek and du Vall 2016) by Niantic, a successful location-based Massively Multiplayer Online Game released in 2012, today available for Android and iOS, and translated over time in sixteen languages. Players divided into two opposing factions—*Enlightened* and *Resistance*—must roam the city to capture *portals* in places of recognized cultural relevance in order to create virtual triangles and control the field.

Niantic, together with Nintendo, released in 2016 *Pokémon Go*, the most renowned location-based augmented-reality game that brought LBMGs to the attention of the wider public worldwide, with more than 750 million downloads and $1,2 billions in revenue in little more than one year from the release according to Apptopia website (Apptopia 2017).

The above-mentioned commercial turn of LBMGs was not the only result of the spread diffusion of smartphones. Companies, research centres and developers followed the path opened by Eric Klopfer at MIT (Klopfer 2008) and exploited the novel availability of GPS-enabled smartphones to create games with an educative goal, but also platforms to democratize the implementation of LBMGs: target audience of these platforms were teachers and educators without any programming skill but keen to involve students in learning experiences through a game.

TaleBlazer (Medlock-Walton 2012) developed by Klopfer's team at the MIT Teacher Education Program is an example. Following the App Inventor style, it allows creating augmented-reality games from scratch, thanks to a user-friendly online editor, and to test them with an Android companion app.

In 2011, the company 7scenes, spin-off of Waag Society, capitalizing on the experience matured with *Frequency 1550*, released a platform for mobile storytelling addressed to non-programmers interested in creating location-based stories and games. Composed by an online editor to create the game, and by an iOS and Android app to test and play the game, the platform was commercialized under the name of *7scenes* and proposed for educational goals as *MLA—Mobile Learning Academy*.

A similar approach is proposed by the *Aris Games*, an open-source platform, made of an online editor and a companion app for iOS that allows to easily create and play LBMGs and interactive stories.

Although the LBMGs created with platforms like those above mentioned are far from commercial ones in terms of complexity of the gameplay and graphical quality, they can perfectly suit for prototyping aims. For example, in the educational context of the School of Design, Politecnico di Milano, *Aris* has been used several times for fast prototyping of persuasive LBMGs (Ackermann and Mariani 2015) and *MLA* platform has been employed to create the 44 LBMGs object of study in this essay.

Looking back at the LBMGs here briefly described, from the turn of the century till today, we can highlight a major difference between commercial and non-commercial ones: commercial LBMGs can be usually played *anytime and anywhere*, in order to widen as much as possible, the audience, while non-commercial LBMGs usually are event games, to be played in a specific place and at a defined time. Being event based is therefore one of the defining characteristics of non-commercial LBMGs, underlined also by Markus Montola who addresses LBMGs as games typically staged at established social events where people can go and play. These games can have different sizes and times: they can vary from a part of a quarter, to the entire city, and they can last for about an hour to a whole day.

2.2   Location Matters

Locating players and providing them with contextual gameplay is at the very core of LBMGs. Accordingly, they can be ascribed to the wide and inclusive group of *location-based services*—LBS—applications integrating geographic location—as spatial coordinates and positioning—with service systems (Schiller and Voisard 2004). In detail, we can define LBS as services able to dynamically update information and contents on the basis of the actual position of users. Therefore, the ways of locating the final user, the player and the precision that can be achieved acquire a great relevance (Maggiorini et al. 2014).

Mobile phones can serve to this aim, but relying only on the capability of mobile devices to locate users may be misleading. First of all, it is necessary to consider the high energy costs of using location sensors and the fact that certain architectures, as city buildings, conceal the GPS signal. In consequence, these two constraints jeopardize the possibility of a continuous and ubiquitous use of localized access (Xanthopoulos and Xinogalos 2016). Secondly, reading the device location can contain errors leading to inaccuracies, as underlined by Xanthopoulos and Xinogalos (2016) who list three reasons why automatic location can flaw: (i) provider trade-offs, (ii) location re-estimation and (iii) varying accuracy. Furthermore, although the commercial LBMGs mostly rely on GPS and A-GPS and, before smartphones diffusion, just on cell positioning, the automatic tracking of players' position is not the unique way of locating them in the playground, being them indoor or outdoor.

Self-reported positioning (Benford et al. 2004), which is asking players to locate themselves in the space, is the most basic approach employed in LBMGs and is usually achieved with several strategies. Players can be prompted to manually locate themselves on the map—i.e. selecting the position on the mobile device screen—or employing elements located in the gamespace, as typing a code that can be retrieved only at a specific place, scanning a QR code placed in the game board, or approaching the mobile phone to a NFC tag. These strategies can be applied both indoor and outdoor and ask for the collaboration of players to carefully locate themselves in the game board by performing specific actions, be them included or not in the gameplay.

Self-reported positioning can also be associated with automatic positioning systems in order to better locate players. The *check-in* is an example: the GPS system approximately locates the user who specifies manually his or her actual position in the vicinity of points of interest—PoIs—such as stores, monuments. *Foursquare* (Cramer et al. 2011; Frith 2013), the renowned location-based and gamified social network, provides an example in this sense.

The Bluetooth beacons are another way of locating mobile devices and consequently their owners within a customizable range, without employing the satellite positioning system. This technology can be employed outdoor, but its potential is best exploited in closed spaces, where the GPS signal can be shielded and highly instable. Nevertheless, Bluetooth beacons do not still assure high accuracy and

stable and precise location of players (Nilsson et al. 2016) and can be comple-
mented by other indoor positioning systems such as the wireless, ultrasonic,
infrared ones (Nuaimi and Kamel 2011).

Whatever the technology chosen to locate players indoor or outdoor, accuracy
issues play a relevant role in LBMGs, since flaws can negatively impact on the
gameplay and strategies to overcome possible inaccuracies must be planned during
the design phase. Most of the games discussed in this book have been designed
following two approaches to manage uncertainty in locating players. On the one
hand, students coupled the automatic positioning system with self-reported positioning and integrated them within the gameplay, and, on the other, they used a
seamful design approach (Chalmers and Maccoll 2003; Chalmers et al. 2005; Broll
and Benford 2005), displaying uncertainty to players by using it as a game element.

Evidently, the two approaches do not exclude each other and can coexist in the
same LBMG: imputing eventual flaws in the GPS signal to the malevolent inter-
ference of an opponent in the game, whoever he or she is, and asking players to
input special codes retrieved in the game board to overcome the malign action is a
clear example of how the two strategies can coexist and be integrated in the
gameplay.

The choice of the most suitable locating system is not only linked to the level of
accuracy needed, to the setting—indoor/outdoor—and to the gameplay. It must
consider also the very nature of the LBMGs. Concurring with this point, Markus
Montola (2011) distinguishes four typologies of location-based games: *physical,
local, global* and *glocal*. Games in the first typology, *physical*, are often characterized as event games, since they require scenography and props, and do not
necessarily include mobile gaming. As a consequence, location is not necessarily
key to the game. The second typology, *local*, includes games strongly context
dependent such as tourist games; while the third, *global*, on the contrary can be
played everywhere, as it happens for *Pokémon Go*. The check-in method—
automatic and self-reported positioning—could be the right choice for local games,
while *global games* should clearly rely on GPS, WLAN or cell positioning. The last
category mixes the previous two, because *glocal* games use the surroundings, but
the content and the gameplay are designed in a flexible manner in order to be easily
relocated. GPS, self-reported positioning and even Bluetooth proximity can suit
with *glocal* games.

Starting from Montola categorization (Montola 2011), we could define the
LBMGs object of study as a hybrid of *physical* and *glocal*, since they are mostly
outdoor event games that markedly impact on the surroundings with scenography
and props. Although they refer to the space wherein they are set with contextual and
coherent mechanics, the content as well as the gameplay has been designed in a
flexible way in order to be easily displaced and rearranged. Consequently, the GPS
location system matched with self-reported positioning was the locating strategy
mostly preferred.

2.3 LBMGs as Pervasive Games

The traditional definition of play as something that happens into well-defined boundaries of time and space and that it is enacted by players who voluntarily accept to conform to rules (Huizinga 1938; Caillois 1958; Salen and Zimmerman 2004) becomes untenable when it comes to LBMGs. The metaphorical idea of a magic circle advanced by Huizinga (1938), reframed by Caillois (1958) and formalized by Salen and Zimmerman (2004) as *magic circle of gameplay*, weakens when referred to games that are played in an undefined urban space, are not constrained by time limits and involve non-players in the gameplay.

Acknowledging the critics moved to the concept of magic circle (Taylor 2006; Malaby 2007; Consalvo 2009) and its defence (Zimmerman 2012; Stenros 2014), we consider it as a powerful metaphor that can define, by contrast, what LBMGs can be: pervasive games, namely games that spatially, temporally and/or socially expand the contractual magic circle of play (Montola et al. 2009).

Pervasive games spatially expand the magic circle when players are taken to uncertain and undedicated areas (Montola et al. 2009), challenging and usually breaking the playground peculiarity of being well defined and constrained. Thus, the hybridization of physical and digital space does not make a LBMG necessarily pervasive. Referring to the categorization of LBMGs proposed by Montola (2011), we could assert that only *global games* are really pervasive, since the digital playground is continuously rebuilt by the system according to player's location. *Pokémon Go* is an example of pervasive LBMG, since it can be played everywhere, in uncertain and undedicated space. On the contrary, *physical games*, characterized by a scenography and props, are mostly event games that happen in a defined space, be it indoor or outdoor. The same could be said regarding *local games*. Nevertheless, When LBMGs are played across the city, they are inherently pervasive, since they take place in undedicated spaces and challenge the use of shared public spaces, changing their role and eventually their appearance.

Games can be considered pervasive also if they expand the temporal boundaries of the magic circle, overcoming the concept of play session towards a more interrelated experience between life and game. In pervasive games, the temporal boundaries of game are uncertain, blurring life and play for the entire duration of the game (Montola et al. 2009). Therefore, LBMGs can be pervasive or not in terms of temporal expansion: games like *Botfighters* are clearly pervasive, since players are always engaged—day and night—while others, such as *PacManhattan*, are played in a well-defined time span.

Finally, games are pervasive if they socially expand the magic circle: the loss of boundaries of time and space can indeed result in the involvement of outsiders in the game, turning bystanders into unaware participants (Montola et al. 2009). This is particularly true for urban LBMGs, since the act of playing can be perceived as something out of context, and consequently involving directly or indirectly passers-by as bystanders and spectators.

As a matter of fact, LBMGs are not necessarily pervasive games but it is also true that, by virtue of their characteristics, they can be easily turned into pervasive games. The fact itself of relying on mobile devices and on the hybridization between the real and the digital worlds makes LBMGs spatially and temporally pervasive: they can turn every space into a game board, and they can be played anytime. In other words, they can impact on the *arena* of play, namely the temporal, spatial or conceptual area that is recognized as ruled for playing and as such culturally acknowledged (Stenros 2014). The social expansion is not necessarily linked to their mobile nature, since the choice of involving or not non-players in the game is up to the designer.

Contextualizing the LBMGs object of study within this theoretical framework, we could maintain their being mostly pervasive, since they spatially expand the magic circle by transforming the university campus into a playground, and socially by frequently involving bystanders into the play activity. However, being designed as event games, they tend not to expand the temporal dimension of the magic circle.

Nevertheless, the pervasive nature of games does not rely only on their technological nature. As Montola and colleagues argue, pervasive games can be "visceral" and tangible experiences play a central role, with important consequences in terms of design (Montola et al. 2009). The materiality of the game experience with players moving in a physical space and doing things for real is therefore key, since it involves players mentally and physically (Montola et al. 2009).

References

Ackermann J, Mariani I (2015) Re-thinking the environment through games. Designing location based mobile games in higher education for environmental awareness. In: Innovation in environmental education: ICT and intergenerational learning. IBIMET-CNR, Firenze, pp 73–78

Akkerman S, Admiraal W, Huizenga J (2009) Storification in history education: a mobile game in and about medieval Amsterdam. Comput Educ 52:449–459

Apptopia (2017) Pokémon GO—app store revenue, download estimates, usage estimates and SDK data. https://apptopia.com/ios/app/1094591345/intelligence. Accessed 3 Nov 2017

Ballagas R, Borchers J (2007) REXplorer: a mobile, pervasive game. In: Von Borries F, Waltz S, Bottger M (eds) Space time play: computer games, architecture and urbanism: the next level. Birkhauser, Boston, pp 366–367

Ballagas R, Kratz S, Borchers J et al (2007) REXplorer: a mobile, pervasive spell-casting game for tourists. CHI '07 extended abstracts on human factors in computing systems. ACM, New York, pp 1929–1934

Ballagas R, Kuntze A, Walz S (2008) Gaming tourism: lessons from evaluating REXplorer, a pervasive game for tourists. In: Indulska J, Patterson DJ, Rodden T, Ott M (eds) Pervasive computing. Springer, Berlin, pp 244–261

Benford S, Seager W, Flintham M et al (2004) The error of our ways: the experience of self-reported position in a location-based game. UbiComp 2004: ubiquitous computing. Springer, Berlin, Heidelberg, pp 70–87

Benford S, Crabtree A, Flintham M et al (2006) Can you see me now? ACM Trans Comput-Hum Interact 13:100–133. https://doi.org/10.1145/1143518.1143522

Broll G, Benford S (2005) Seamful design for location-based mobile games. Entertainment computing—ICEC 2005. Springer, Berlin, Heidelberg, pp 155–166

Caillois R (1958) Man, play and games. Free Press, Chicago

Chalmers M, Maccoll I (2003) Seamful and seamless design in ubiquitous computing. In: Proceedings of workshop at the crossroads: the interaction of HCI and systems issues in UbiComp

Chalmers M, Bell M, Brown B, et al (2005) Gaming on the edge: using seams in Ubicomp games. In: Proceedings of the 2005 ACM SIGCHI international conference on advances in computer entertainment technology. ACM, New York, NY, USA, pp 306–309

comScore (2011) The comScore 2010 mobile year in review

Consalvo M (2009) There is no magic circle. Games Cult 4:408–417. https://doi.org/10.1177/1555412009343575

Cramer H, Rost M, Holmquist LE (2011) Performing a check-in: emerging practices, norms and "conflicts" in location-sharing using foursquare. In: Proceedings of the 13th international conference on human computer interaction with mobile devices and services. ACM, New York, NY, USA, pp 57–66

de Souza e Silva A, Delacruz GC (2006) Hybrid reality games reframed: potential uses in educational contexts. Games Cult 1:231–251. https://doi.org/10.1177/1555412006290443

Flintham M, Benford S, Anastasi R et al (2003) Where on-line meets on the streets: experiences with mobile mixed reality games. In: Proceedings of the SIGCHI conference on human factors in computing systems. ACM, New York, NY, USA, pp 569–576

Frith J (2013) Turning life into a game: foursquare, gamification, and personal mobility. Mob Media Commun 1:248–262. https://doi.org/10.1177/2050157912474811

Huizinga J (1938) Homo Ludens, 2002 edition. Giulio Einaudi Editore, Torino, Italy

Huizenga J, Admiraal W, Akkerman S, Dam GT (2009) Learning history by playing a mobile city game. J Comput Assist Learn 332–344

Jacob J, da Silva H, Coelho A, Rodrigues R (2012) Towards Location-based augmented reality games. Procedia Comput Sci 15:318–319. https://doi.org/10.1016/j.procs.2012.10.093

Kelley MA (2006) Local treasures: geocaching across America. Center for American Places, Santa Fe, Staunton

Klopfer E (2008) Augmented learning: research and design of mobile educational games. MIT Press, Cambridge, MA

Klopfer E, Squire K, Jenkins H (2002) Environmental detectives: PDAs as a window into a virtual simulated world. In: Proceedings of IEEE international workshop on wireless and mobile technologies in education, pp 95–98

Klopfer E, Perry J, Squire K et al (2005) Mystery at the museum: a collaborative game for museum education. In: Proceedings of the 2005 conference on computer support for collaborative learning: learning 2005: the next 10 years! International society of the learning sciences, Taipei, Taiwan, pp 316–320

Maggiorini D, Quadri C, Ripamonti LA (2014) Opportunistic mobile games using public transportation systems: a deployability study. Multimedia Syst J 20(5):545–562

Majorek M, du Vall M (2016) Ingress: an example of a new dimension in entertainment. Games Cult 11:667–689. https://doi.org/10.1177/1555412015575833

Malaby TM (2007) Beyond play: a new approach to games. Games Cult 2:95–113

Medlock-Walton MP (2012) TaleBlazer: a platform for creating multiplayer location based games. Thesis, Massachusetts Institute of Technology

Milgram P, Kishino F (1994) A taxonomy of mixed reality visual displays. IEICE Trans Inf Syst 12:1321–1329

Montola M (2011) A ludological view on the pervasive mixed-reality game research paradigm. Pers Ubiquitous Comput 15:3–12. https://doi.org/10.1007/s00779-010-0307-7

Montola M, Stenros J, Waern A (2009) Pervasive games. Experiences on the boundary between life and play. Morgan Kaufmann Publishers, Burlington, MA

Nilsson T, Blackwell A, Hogsden C, Scruton D (2016) Ghosts! A location-based bluetooth LE mobile game for museum exploration

Nuaimi KA, Kamel H (2011) A survey of indoor positioning systems and algorithms. In: 2011 international conference on innovations in information technology, pp 185–190

O'Hara K (2008) Understanding geocaching practices and motivations. In: Proceedings of the SIGCHI conference on human factors in computing systems. ACM, New York, NY, USA, pp 1177–1186

Raessens J (2007) Playing history. Reflections on mobile and location-based learning. Waxmann Verlag, Munster

Salen K, Zimmerman E (2004) The rules of play: games design fundamentals. MIT Press, Cambridge, MA

Schiller J, Voisard A (2004) Location-based services. Morgan Kaufmann, San Francisco

Squire KD, Jan M, Matthews J et al (2007) Wherever you go, there you are: place-based augmented reality games for learning

Stenros J (2014) In defence of a magic circle: the social, mental and cultural boundaries of play. Trans Digit Games Res Assoc 1. https://doi.org/10.26503/todigra.v1i2.10

Taylor TL (2006) Play between worlds: exploring online game culture. The MIT Press, Cambridge, Mass

Xanthopoulos S, Xinogalos S (2016) A review on location based services for mobile games. In: Proceedings of the 20th Pan-Hellenic conference on informatics. ACM, New York, NY, USA, pp 28:1–28:6

Zimmerman E (2012) Jerked around by the magic circle—clearing the air ten years later. Gamasutra

Chapter 3
Informal Learning Between Design and Play

Abstract This chapter addresses the practice of designing and the activity of playing as matters of learning out of formal education activity. Theoretically, we frame LBMGs learning outcomes in the fields of mobile learning, situated learning and game-based learning, focusing then on the process of meaning-making we observed in our didactic activity.

Keywords Informal Learning · Game-Based Learning · Learning Outcomes Game Experience

3.1 Theoretical Premises

This chapter addresses the practice of designing games and the activity of playing as means of learning, looking at LBMGs for their ability to stimulate active engagement and situated learning (Lave and Wenger 1991). Relying on theoretical premises pertaining to the fields of mobile learning, situated learning and game-based learning, and on the didactic experience conducted over three years, we propose a novel perspective on LBMGs, intended as activators of social change, transformation and meaning-making. The point of view is not unique, but broad and inclusive. LBMGs are here analysed as communication tools to be designed and played, as practices able to impact on both designers and players, being the first involved in the act of crafting the game, and the second involved in the play activity. The filter is that of informal learning, namely that kind of learning that happens out of learner's will, and out of formal education settings such as schools and often in adulthood (Merriam and Caffarella 1999). Discussing the potentials of LBMGs in an informal education setting means addressing broad and diversified fields of research such as mobile learning (m-learning), situated learning, and game-based learning (GBL), since games—and mobile-supported located games in particular—are complex communication systems which establish multifaceted relations between players, the context of play, the technology, and other players.

D. Spallazzo and I. Mariani, *Location-Based Mobile Games*,
PoliMI SpringerBriefs, https://doi.org/10.1007/978-3-319-75256-3_3

Looking at LBMGs as mobile-supported systems, it is indeed necessary to address issues related to mobile learning (m-learning), a very diversified and multidisciplinary field still in need of a unique, shared definition. Early definitions of m-learning intend dialogue as the focal process involved in a learning activity, able to activate negotiation of meanings as well as the creation of a stable even if transient interpretation of the world (Sharples et al. 2005). Dialogue is pointed by Sharples and colleagues as a process happening with and through technology: they make no distinction between humans and machines, and highlight how their definition lacks in recognizing the moral and social worth of human beings in respect to technology.

Therefore, the social dimension—specifically investigated in this chapter, although through a different perspective—is not considered in this early definitional attempt, but it is included by Klopfer et al. (2002) as one of the four key factors of mobile and context-aware technology, together with portability, context sensitivity and individuality. The authors, by listing social interactivity and individuality as focal characteristics of this kind of technology, acknowledge the seemingly conflicting nature of mobile technology, at the same time rigorously personal but able to trigger social interaction and sharing.

Similar considerations can be found in Naismith et al. (2004) that expand the reasoning by listing five key issues to be considered in the design of mobile learning experience: context, mobility, learning overtime, informality and ownership. Interpreting this list, mobile learning appears to be an informal process happening overtime and in mobility, and strongly related to the context.

Mobility is therefore a key property characterizing the interaction between people and technology, interpreted by Kukulska-Hulme et al. (2009) in a broader sense, as a combined experience of retrieving data through different media. The result is an extended notion of mobility that appears to be multifaceted in five intertwined aspects: (i) mobility in *physical space* referring to people on the move; (ii) mobility of technology that refers to both the *portability* of the devices and the possibility to transfer attention across devices; (iii) mobility in *conceptual space*, referring to mobility between a concept or a topic to another one; (iv) mobility in *social* space that refers to the different contexts the learners perform within; (v) *learning dispersed overtime* that describes learning as a combined experience that happens through different media and across time. These five key features of mobility acquire further relevance once applied to LBMGs, since they seem to fulfil them all: players indeed move in the urban space browsing the surroundings with the help of a portable device, looking for hints and contents, and potentially they can socially engage with other players and non-players. Therefore, mobile learning appears as a process of informal acquisition of knowledge, experience and awareness while in mobility and enhanced by personal and public technology (Kukulska-Hulme et al. 2009).

The relation with the context, being it perceived by learners/players through contextual information provided by the mobile device or directly exploring the real world, is at the basis of the situated learning paradigm mainly developed by Lave and Wenger (1991). It claims that learning is not merely a personal acquisition of knowledge but rather a complex process characterized by social participation (Naismith et al. 2004) that requires interaction and collaboration. Moreover, the fact of providing information, and consequently knowledge, in the real context acquires great relevance (O'Malley et al. 2003).

Relying on this approach, informal learning appears to be unintentional and gained through progressive involvement in a *community of practice*. Following the concept of *legitimate peripheral participation*, by moving from the periphery of the community to its centre, learners become more active and engaged in the culture, until assuming the role of experts (Lave and Wenger 1991; O'Malley et al. 2003).

Gaining knowledge presented in authentic contexts and with learners involved in a community, a social group, is therefore at the basis of the situated learning, it seems to well fit with mobile-supported experiences—such as LBMGs—since mobile technology can be a powerful support to situated learning with its portability, context-awareness and connectivity.

Enhancing and augmenting the learning experience through mobile technology is matter of discussion in Klopfer's essay *Augmented Learning* (Klopfer 2008) that argues about the potential of mobile-supported learning in real-world contexts. By matching the potential of mobile games to enhance engagement and learning out of formal education activities with the possibility of embedding learning in authentic environments through location-based technologies (Huizenga et al. 2009), he theorizes the use of LBMGs as powerful means of informal education as well as of enforcement of formal education.

This reasoning finds further reference in Prensky's, who sustains that the mixing of fun and interactive entertainment with serious learning is also at the basis of digital game-based learning; a combination pointed by the author as a way of addressing contemporary learners both in formal and informal settings (Prensky 2001).

By playing digital games, players/learners can acquire not only ever-increasing skills, as suggested by Gee (2003, 2004), but they get acknowledged and educated about the specific topics that are subjects of these games. Embracing this point of view, LBMGs appear to be effective in triggering active engagement and situated learning (Lave and Wenger 1991). They possess indeed the ability to match the playfulness of games with the richness of contextual contents, and to transport players/learners into a hybrid world, between reality and virtuality (Klopfer et al. 2002; Klopfer 2008).

How to balance the engaging and entertaining side with learning in digital games that have an educative aim is a debated issue. The relevant role of the matter is underlined by Avouris and Yiannoutsou (2012) who proposed to use it as basis for their classification of LBMGs, that is, distinguished in three categories according to their final aim: (i) *ludic*, whose aim is to make players enjoy; (ii) *pedagogic* that includes LBMGs overtly developed for learning and (iii) *hybrid* that collects games with both the aforementioned aims—entertain educating. Learning can happen by playing games pertaining to all the three categories, but while it is an explicit goal for the *pedagogic* and the *hybrid* categories—that we can call *serious game*

(Ulicsak and Wright 2010)—it is a side effect for the *ludic* category (Avouris and Yiannoutsou 2012).

Players can be therefore transported into a hybrid world, created by designers to engage them and—voluntarily or eventually—transfer meaning related to a great variety of topics that LBMGs can cover.

3.2   LBMGs and Learning Outcomes

The literature about mobile, situated and game-based learning summarized above focused specifically on the learning outcomes of users/learners. *Learning by playing/experiencing* is the attitude described until now that intends mobile experiences as well as LBMGs as systems to transfer meanings.

During the three-years-long didactic experience conducted with design students crafting and playing LBMGs, we explored them as educational tools from diverse points of view, identifying the learning outcomes in three categories here summarized:

- (L1) *learning to design* for mobile experiences: formal learning within
- design practice;
- (L2) *learning by designing* games: formal learning about the design objective
- of the course and informal learning about specific topics covered by each game;
- (L3) *learning by playing* games: informal learning about specific topics.

The first approach (L1) considers LBMGs as significant mobile experiences, whose design is matter of the course during which the mobile games have been conceived and tested. They are indeed the result of a formal didactic activity within the field of design, and interaction design in particular, aimed at teaching undergraduate students how to design for mobile experiences with a holistic approach. Students are asked to consider not only the sole interaction with mobile devices, but also its development overtime, the context in which it happens, and, above all, the meaning of the interaction. Shifting the attention of future designers from the implementation of new technologies towards the design of the resulting experiences (Hassenzahl et al. 2013) is indeed focal in the course approach.

Therefore, designing LBMGs is first of all a formal education activity (L1) carried out by mixing theoretical lectures and hands-on experiences. This activity can be targeted as *learning to design*, since students acquire skills to conceive meaningful mobile experiences by designing LBMGs.

Devising a game requires also specific knowledge in the discipline of game design, provided to students in the form of short lectures. Therefore, the act of designing games also becomes a means of formal education about game design, since students should master the basics of the discipline to design coherent, working and engaging games. This activity configures a second category of learning, classified in the list above as *learning by designing* (L2)—namely a formal learning activity about game design. Designing a game requires also knowledge of the issue the game deals with, in order to create a meaningful fictional world, relevant tasks, and coherent game mechanics. Consequently, *learning by designing* means learning how to design games—learning by doing—but also gaining both a wide and narrow knowledge about the topic. Reading up on the topic, organizing the information, translating them into the game and having a point of view on the debated issues are activities that, despite not central to the formal education activity, can provide learning as side effect. This kind of learning (L2) affects mainly the design team that makes use of design skills to convey the acquired knowledge, in the shape of a game experience, to players that will—hopefully—learn by playing (L3).

The third category of learning triggered, as dealt in the didactic experience, is *learning by playing* (L3) that focuses on the ability of games to inform players about topics while engaged in the play activity.

In this approach, games become triggers of conversation and debate between senders and receivers, namely designers and players, since they foster negotiations of meanings conveyed through the game and the play experience it provokes (Sicart 2011). The meaning is the message that game designers, as senders, intend to transmit, and it emerges as a result of the player's interpretation (see Chap. 4).

The three categories of learning discussed above—to design/by designing/by playing—underlie both formal and informal learning activity, being the first mostly related to the design activity, and the second to the topic addressed by the game. Although the formal learning approach is of great interest and may highly contribute to the discipline of design, the slant of this short volume—that we recall is not intended to be a handbook of LBMGs—privileges a focus on the informal learning approach. Consequently, the attention is mainly on L2 and L3, which are considered as entangled, being the study grounded into the design discipline. As a matter of fact, our practice implies both translating topics into games (L2) and analysing how players make sense and meaning out of them (L3).

Flanagan and Nissenbaum in their essay *Values at Play in Digital Games* (2014) keep a similar approach. They address the attitude of learning by (L2) designing and (L3) playing games, discussing digital games as means for conveying and integrating moral and political values. Through their research, and benefitting from the workshops conducted during the *Values at Play* project, as well as the *Grow-a-Game* brainstorming tool, the authors dig into how game-based systems communicate ideas and incorporate human values. In our study, their systematic approach has been adopted and further extended to LBMGs and their design process, challenging design students with the constraints of mobile technology and real world, and focusing in particular on the aspect of meaning-making as a consequence of the design and play activity.

3.3 Learning by Designing and Playing

In our didactic experience, we considered games as (i) systems to convey meanings, as (ii) methods to reflect on complex issues and multifaceted perspectives into in-context experiences, and as (iii) tools of investigation to better comprehend the translation from theory to practice. In particular, we asked design students to explore games and their dark side (Antonacci 2012; Sicart 2014), challenging unfriendly issues as socio-cultural taboos and societal wicked problems (Sicart 2011).

Students created LBMGs centred on subjects of apparently debatable ludic interest, often imbued with negative values and characterized by a spread tendency to be avoided in daily conversations. Our academic interest was indeed to analyse games as triggers of reflection on societal issues, testing the ability that Juul (2013) ascribes to games, that of enabling players to deliberately explore morally untenable choices, and make negative emotions and feelings arise.

Through games, both designers and players are enabled to argue and discuss on topics that in daily life may be considered unkind, grave or harsh, dealing such issues through a novel lens. In so doing, designers and players are made aware of the issues and are called to take a stand on them: designers are challenged to convey the meaning from a specific point of view, and players to share the position or not. Furthermore, the design team is asked to gather documents and information on the issue to cover, to analyse and organize them in order to share a common knowledge on the topic inside the team, and to translate this knowledge into the game.

During our didactic activity, we noticed that design teams must acquire both a wide and narrow view on the topic to successfully craft the games in their entirety, considering not only the message to be conveyed but also creating coherent and consistent game mechanics, bearing in mind the actions they aim to trigger. The necessity to gain both wide and deep knowledge resulted essential when dealing with issues such as illness. An example comes from the game *The Lost Papyrus* (Benedetti et al. 2015), which addresses the Alzheimer's disease aiming at informing players about the effect of this illness has on the everyday activities of sick people and of those who live with them.

Given such a delicate topic, the design team decided to address it through a metaphoric narration, dropping players in the early XX century in the shoes of an expert archaeologist of Egyptian culture and his three valiant assistants (Fig. 3.1). The choice of the metaphor stemmed from an attentive study of the common symptoms of the disease and of its degeneration overtime, often characterized by problems with language, disorientation, difficulty to set actions in the correct order and reduced mobility.

The metaphor of the Ancient Egypt allowed to easily turn these symptoms into game mechanics, such as translating and encrypting hieroglyphics, finding the way out of tricky passages or ordering actions. The fact of expressing the sense or consequences of the symptoms into games mechanics, as well as and the creation of a metaphorical story that integrates and communicates them, are clear evidences that an informal learning happened inside the design team. While learning to craft LBMGs (formal learning) designers acquired knowledge on the issue that brought them to cover it with the due respect, in a metaphorical way, and disclosing the intended message to players only at the end of game. This final disclosure acts as a final revelation that pushes players to re-read the game, adding a further level of meaning to the activities done, and grasps the real messages embedded in the game

Fig. 3.1 Players of the *The Lost Papyrus* (Benedetti et al. 2015)

(see Chap. 6). This choice was widely discussed inside the team, confronting the acquired information with personal experiences, and debating different points of view. As a matter of fact, discussion and confrontation were at the basis of a learning process that brought the design team to share a common view on a tactful topic. At the same time, they necessarily adopted a narrow view, studying the most common symptoms, their degeneration, and how they are usually perceived by sick people and their families. The significant number of documents read and discussed and information filtered and selected by the students, resulted in an unintentional but thoughtful knowledge of the disease.

Other games developed in the augmented reality and mobile experience course chose a different approach to deal with difficult topics: *The Infection* (Bassanese et al. 2015) aimed at sensitizing players on the control, prevention and treatment of sexually transmitted diseases (STDs) keeping an ironical tone. In the game, players wear the shoes of four dangerous but common viruses—gonorrhoea, syphilis, human papillomavirus and chlamydia—interpreted as underworld bosses whose aim is to infect the city. In the role of criminals, players must face four challenges that deal with issues such as hygiene, protection, infection and symptoms control (Fig. 3.2). As it happened for the aforementioned game *The Lost Papyrus*, the design team collected as much information as possible about venereal diseases in order to structure a coherent and meaningful LBMG, imagining the fictional world and the game mechanics that mostly fit their communication purposes. The choice of an ironical tone of voice to deal with such a sensitive and somehow taboo issue is emblematic of the process of informal learning. The design team needed to acquire

Fig. 3.2 Players of the *The Infection* (Bassanese et al. 2015) personifying the underworld bosses—Gon O'Rock, Calamity Clam, El Papito, Don Sifilio

both a wide and specific knowledge about the topic to lightly and sarcastically discuss the subject without diminishing its impact on players. At the same time, to match the pleasantness of the game with both the instructional power and ironical but not trivial tone, the act of translating aspects of venereal diseases such as prevention, infection or symptoms control into game mechanics asked for a great work of design and progressive refinement.

The process of translating information, concepts, processes and knowledge into a story, a fictional world and game mechanics are therefore at the basis of the process of informal learning, since it requires collection of data and interpretation. As a matter of fact, each game envisioned by the students is the result of a design process but at the same time it embeds meaning and acts as a communication system, configuring as evidence of a learning process.

The games realized and played by the students are clear proofs of the learning activity the design team went through. However, the testimonies of learning outcomes from players are much more difficult to be collected.

Quantifying the learning outcome of a mobile-supported experience is never an easy task, especially when it comes to persuasive LBMGs, whose aim is not providing players with notions, but sensitizing them about socially relevant issues. Structured questionnaires filled in by players at the end of the play experience as well as informal interviews may shed light on how they perceived the game and were touched by it, but cannot account on real transformation of players' attitude towards the subject.

In our study, we observed that frequently players during the play activity were empowered to think and act according to the role they were assigned, and, in so doing, to embrace and discuss the point of view proposed by the design team (Mariani and Spallazzo 2016). Very often players reported that games invited them to question their morality and ethics, acknowledging games as something more than simple tools intended to make unpleasant things enjoyable or to transfer basic concepts. In other words, the designed games provided players with a stimulus to discuss and question delicate issues, potentially able to affect and change their point of view. As a matter of fact, thanks to their gameplay, narratives, challenges, and in-game choices these games involved players in processes of layered understanding: (i) of the world wherein they were transported, (ii) of their rules as well as (iii) of the point of view of the characters/roles they were impersonating. In so doing, they were asked to comprehend the sense of the game and of the in-game mechanics, translating step-by-step the message of the game.

Moreover, it emerged that the awareness of being into a safe representation, contributed to framing the game as a space of openness, where the player can self-challenge by questioning her own attitudes, meanings and frames. This process of re-examination and re-attribution of sense are on the ground of the learning activity (Anolli 2010). Especially, the encounter of failure within a structure characterized by motivation, as the game, nurture recursive learning (Mitgutsch 2012). Since the cost of failing when playing is low, players have a propensity for putting themselves to the test, becoming even emotionally capable of opening to different perspectives, trying out ideologies that are not their usual ones.

These considerations, emerged by questionnaires, direct observation and informal interviews (Mariani and Spallazzo 2016) do not account for permanent changes in the mindset of players: longitudinal studies, clearly out of reach for a didactic activity, are needed in order to verify significant changes and quantifiable learning outcomes. Nevertheless, the analysis of data gleaned during our study highlighted some aspects that seem to impact on the quality of the game experience and the effectiveness in communicating the message of the game.

3.4 Considerations from the Empirical Study

We observed a noticeable direct relation between the capacity of games to immerse players in the story, and their ability to convey the main message, being the first aspect mostly related to the appeal of the graphics of both the physical materials and the visual on the smartphone app; while the second relies more on the consistency of all the game aspects (story, game mechanics, game kits) with the fictional world. A pleasant, flowing and coherent game experience, supported by well-designed game kits and apps, results to be functional in favouring immersion and enjoyment and in so doing conveying the main message.

Well-designed and coherent fictional worlds seem to foster a condition of both separation and openness, able to momentarily detach players from their usual way to act, augmenting their responsiveness towards the topics addressed by the game. Assigning roles to players, and asking them to wear the shoes of characters in a story and act accordingly, allowed them to think out of their box and to question their morality and ethics (see Chap. 5; Mariani 2016).

The identification with a character and the immersion of players in the fictional world was increased using physical objects and deploying a strong connection between the fictional world and the urban setting in which the game took place. In this sense, we noticed that the interaction with physical objects and with the real environment added a sense of physicality and realism to the digital experience usually provided by mobile games. The mobile experience, augmented by physical interactions, resulted more remarkable and therefore successful in conveying the message of designers.

Another relevant issue emerged from our study is the role of the end of the game in providing players with an opportunity to look back at the play experience from a different point of view (see Sect. 6.2). In our didactic experience, we questioned the tacit rule that games must end with a win or a loss, with the winning condition as the ideal one (Ruggiero and Becker 2015). As a matter of fact, our analysis shows that failure as possible outcome of the game can be an effective means to trigger meaning-making. In other words, designers made explicit use of productive failure or loss, citing Kapur (2008) and extending his reasoning to games. In particular failure can convey the underlying message and stimulate a reflection on the game experience, transforming the end of the game in a moment fraught with meaning.

In our didactic activity, we asked students to work in team to design LBMGs to be played in groups of three to four players. The aim was to study, on the one side, how designers collaborate and elaborate meaning out of a design activity and, on the other, to observe the social dynamics inside the group of players and how they engaged in a common effort to complete the game (see Chap. 4).

Looking at the activity of groups of players from the point of view of learning, we observed that they shared experiences, skills and resources, engaging in a common learning process. Players employed face-to-face conversation as the main means of collaborative elaboration of the information: they discussed together the right answer to a question, the interpretation of the map, the direction to keep, the roles within the group and so on. Referring to the conversation theory (Pask 1976) players went through a learning process in the form of conversation between different systems of knowledge, being them players with their cultural background, or players wearing a different role and consequently proposing views of the world potentially different from their own. In other words, through conversation and discussion, players may question their mental maps and their model of the world, that, according to Argyris and Schön (1974), direct people's actions and modify the way people evaluate the cause-effect chain.

According to O'Malley and colleagues, learning happens when people share their description of the world and through conversation they come to a shared understanding of it (O'Malley et al. 2003). Translating this concept in the game experience observed in our study, players were asked to share firstly a common understanding of the game fictional world, and secondly to discuss together the meaning of the actions performed and of the entire game experience. They did it by expressing opinions, describing what they observed, and interpreting it in relation to their personal experience.

The social nature of play, already underlined by Huizinga in his seminal work (1938), and the resulting social interaction may indeed empower the learning process. Despite not regarding the field of games, the reflections of Paris (1997) about the role of sociality in cultural learning shed light on the reasons why social interaction may enhance the learning experience. He lists five benefits of social learning: people stimulate each other's imagination and negotiate meaning; the shared goal enhances motivation; there are social supports for learning; people learn through observation, and the companions provide a means of confrontation (Paris 1997).

These points are also echoed in Siemens' connectivism (Siemens 2005) which links an efficacious learning with the ability to build a wide personal network. According to the author, we can no longer acquire all the knowledge we need personally, since in many fields the life of knowledge is now measured in months and years. Knowledge and competences are therefore derived forming connections and a personal network that feeds into organizations and institutions, which feed-back to the network and to the individual (Siemens 2005). Social engagement and social interaction in LBMGs appear therefore to be a powerful trigger of meaning-making, stimulating players to collaborate and share information during the play activity. A comprehensive discussion about the social dimension of play in LBMGs is presented in the next chapter (see Chap. 4).

References

Anolli L (2010) La sfida della mente multiculturale. Nuove forme di convivenza. Cortina Raffaello, Milan, Italy

Antonacci F (2012) Puer Ludens. Antimanuale per poeti, funamboli e guerrieri. FrancoAngeli, Milan, Italy

Argyris C, Schön DA (1974) Theory in practice: increasing professional effectiveness. Jossey-Bass, San Francisco, CA

Avouris N, Yiannoutsou N (2012) A review of mobile location-based games for learning across physical and virtual spaces. JUCS—J Univers Comput Sci 2120–2142

Bassanese G, Bonfarnuzzo L, Pham C, Redana F (2015) The Infection [LBMG]. Politecnico di Milano, School of Design, Milan, Italy

Benedetti A, De Marco A, Franco Conesa CM, Piatti J (2015) The Lost Papyrus [LBMG]. Politecnico di Milano, School of Design, Milan, Italy

Flanagan M, Nissenbaum H (2014) Values at play in digital games, Reprint edn. MIT Press, Cambridge, MA

Gee JP (2003) What video games have to teach us about learning and literacy. Palgrave Macmillan, New York, NY

Gee JP (2004) Situated language and learning: a critique of traditional schooling, 1st edn. Routledge, New York, NY

Hassenzahl M, Eckoldt K, Diefenbach S et al (2013) Designing moments of meaning and pleasure. Experience design and happiness. Int J Des 7:21–31

Huizenga J, Admiraal W, Akkerman S, Dam GT (2009) Learning history by playing a mobile city game. J Comput Assist Learn 332–344

Huizinga J (1938) Homo Ludens, 2002 edition. Giulio Einaudi Editore, Torino, Italy

Juul J (2013) The art of failure: an essay on the pain of playing video games, Reprint edn. MIT Press, Cambridge, MA

Kapur M (2008) Productive failure. Cogn Instr 26:379–424

Klopfer E, Squire K, Jenkins H (2002) Environmental detectives: PDAs as a window into a virtual simulated world. In: Proceedings IEEE international workshop on wireless and mobile technologies in education. IEEE computer society, pp 95–98

Klopfer E (2008) Augmented learning: research and design of mobile educational games. MIT Press, Cambridge, MA

Kukulska-Hulme A, Sharples M, Milrad M et al (2009) Innovation in mobile learning: a european perspective. Int J Mob Blended Learn 1:13–35

Lave J, Wenger E (1991) Situated learning: legitimate peripheral participation. Cambridge University Press, Cambridge, MA

Mariani I (2016) Meaningful negative experiences within games for social change. Designing and analysing games as persuasive communication systems. Dissertation, Politecnico di Milano

Mariani I, Spallazzo D (2016) Empowering games. Meaning making by designing and playing location based mobile games. ID&A Int Des Archit 28:12–33

Merriam S, Caffarella R (1999) Learning in adulthood: a comprehensive guide. Jossey-Bass, San Francisco, CA

Mitgutsch K (2012) Learning through play—a delicate matter: experience-based recursive learning in computer games. In: Fromme J, Unger A (eds) Computer games and new media cultures. Springer, Netherlands, Dordrecht, pp 571–584

Naismith L, Lonsdale P, Vavoula GN, Sharples M (2004) Literature review in mobile technologies and learning. NESTA FutureLab, Bristol, UK

O'Malley C, Vavoula G, Glew JP et al (2003) MOBIlearn. WP 4—guidelines for learning/teaching/tutoring in a mobile environment

Paris SG (1997) Situated motivation and informal learning. J Museum Educ 22:22–27

Pask G (1976) Conversation theory: applications in education and epistemology. Elsevier, Amsterdam/Oxford/New York

Prensky M (2001) Digital game-based learning. McGraw-Hill, New York, NY

Ruggiero D, Becker K (2015) Games you can't win. Comput Games J 4:169–186. https://doi.org/10.1007/s40869-015-0013-9

Sharples M, Taylor J, Vavoula G (2005) Towards a theory of mobile learning. Proc mLearn 1:1–9

Sicart M (2011) Against procedurality. Game Stud 11

Sicart M (2014) Play matters. MIT Press, Cambridge, MA

Siemens G (2005) Connectivism: a learning theory for the digital age. Int J Instr Technol Distance Learn 2:3–10

Ulicsak M, Wright M (2010) Games in education: serious games. NESTA FutureLab, Bristol, UK

Chapter 4
The Social Dimension of Located Play

Abstract The chapter addresses LBMGs as collaborative activities. Relying on theoretical assumptions and inferences drawn from practice-based research, it interprets LBMGs as triggers of social engagement within the group of players and with non-players. It advances design hypotheses on the social configuration of players during the game experience, analysing how they can impact on the activity and on players' perception.

Keywords Social Configuration · Social Engagement · Strangers

By virtue of their nature, partly digital and partly physical, LBMGs allow designers to conceive and design games characterized by rich social interactions, engaging players in a multitude of varied experiences. As a matter of fact, in-game social interaction may occur between players, and/or between players and non-players, and its dynamics depend on the typology of game as well as on its design. Urban LBMGs, in particular, have the potential of involving players in rich, in-person social engagement with those who they may encounter, being them other players and non-players: a peculiarity common to urban games, which, on the contrary, do not usually foster technology-mediated social engagement. LBMGs, like video games, may allow players to dialogue and have social relationships with other players, ranging from being more or less mediated by the technological device (e.g. chat or voice chat, instant messages, calls), to directly sharing the play activity with other players (in-person play activities as challenges, quests, and so on that involve more than a player in the meanwhile).

The issue of the social configuration of players during the play experience is not secondary; on the opposite, it acquires a great relevance for LBMGs designers, who can range from proving players with a solo experience, to making them interact with unknown people. In other words, designers can choose between complete anonymity and affiliation, between working inside or outside of the magic circle.

When designed exploiting the cultural capital of play in public spaces (Sicart 2017), taking advantage of the potentialities of creating partially technology-mediated game experiences, LBMGs offer designers the opportunity to

D. Spallazzo and I. Mariani, *Location-Based Mobile Games*, PoliMI SpringerBriefs, https://doi.org/10.1007/978-3-319-75256-3_4

engage players in a play experience characterized by different social configurations. For example, consciously embedding the way of playing typical of urban games, they can fully exploit in-person social contacts as well as the context wherein they take place. As a matter of fact, designers can provide players with a personal play activity or, on the contrary, require the continuous collaboration of a team of players to proceed. Other games could ask a team to split in order to complete quests, or players to socialize with non-players and even unknown persons, namely strangers (Simmel 1950).

Despite not conceived for the game design field, the theoretical framework proposed by Debenedetti (2003) for describing the models of social appropriation of an exhibit space can be easily translated into LBMGs realm to describe the social configuration of players. The *together-alone* or *accompanied-anonymous* framework is discussed by the author through a semiotic square structured in four modes of social configurations (Fig. 4.1): together, alone, not alone and not together (Debenedetti 2003).

The square *together* may portray a complete sharing of the play experience, with a team of players that follow the same path, share every choice and stay together from the beginning to the end of the game. We can define this social configuration with the term affiliation, that is typical of LBMGs that assign different roles, and consequently abilities, to players. The square *alone*, on the contrary, exemplifies a solo experience, characterized by a single player who faces the entire play experience alone without interacting with other players or non-players, in complete anonymity. In the schema, the two squares are placed in a position of relative contradiction, being the first on the side of a social play experience and the second on that of a solo experience.

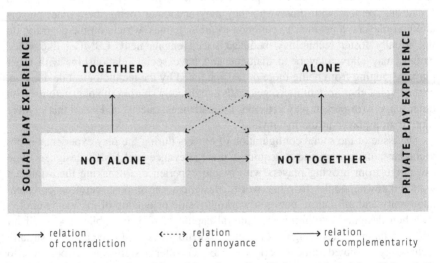

Fig. 4.1 Semiotic square of possible social configurations of players. Adapted from Debenedetti (2003)

The two squares *together* and *alone* are characterized by static social configurations that do not change over time: in team games, the group never splits—together—while in solo games, the players do not interact with other humans, neither in person, nor in a technology-mediated manner.

The other two squares, instead, portray dynamic social configurations and are characterized by the passage from affiliation to anonymity—*not together*—and vice versa—*not alone* (Debenedetti 2003). In relation of complementarity with the square *alone* and of contradiction with the square *not alone*, the condition of being *not together* describes a particular kind of gameplay that allows or requires a team of players to split during the play experience. A condition that can be obtained by letting players voluntarily use verbal, technological or even physical barriers to social interaction, or it can be foreseen by designers in the gameplay.

In contradiction with the condition of *not together*, there is the *not alone*, which describes the opposite behaviour, namely the pursuit of social contact of a solo player, who decides or has to engage socially with other players and/or non-players. As a matter of fact, the change in the social configuration from *alone* to *not alone* can be triggered by players themselves, who decide to exit their anonymity and engage socially with other human beings, being them players or non-players, or it can be necessary to proceed in the game, because defined by the designer in the ideation of the gameplay. The act of engaging with non-players marks the social expansion of the magic circle and formally the passage to a pervasive LBMG.

The theoretical framework here discussed, despite originally designed for another kind of experience, namely a museum visit experience (Debenedetti 2003), can vividly describe how players can be socially configured while playing a LBMGs and sensitize designers in taking into account the social dimension of play. In particular, it can help in ideating a gameplay that dynamically changes the social configuration of players, meaningfully favouring the temporary separation of a team of players—*not together*—ore the social engagement of players with other players or non-players—*not alone*.

4.1 Inside the Magic Circle: Social Configuration of Players

As said, LBMGs are not necessarily pervasive and, even if designed as urban LBMGs, they do not necessarily expand the magic circle socially (Montola et al. 2009). Consequently, it is worth analysing the social configuration of players within the magic circle, studying how they can be socially engaged while keeping inside its safe borders. The above-mentioned *accompanied-anonymous* framework (Debenedetti 2003) offers a good starting point to break down the social configuration of LBMGs players according to the four modes—*alone, together, not together* and *not alone*—and analyse their implications.

Commercial LBMGs very often provide users with play experiences that can be classified within the *alone* social configuration. Games such as *Botfighters*, *Ingress* and *Pokémon Go*, to name the most renowned, are single player games that ask to roam the city in order to fight with other players, subtract territory from the opposing faction or capture digital characters. The players of this kind of LBMGs are not required to engage socially with other players or non-players, but their action has clearly an impact on other players who can be defeated in a fight or loose territories of the city. Nevertheless, despite designed to be single player games, these successful LBMGs frequently witnessed spontaneous groupings of players, who joined in teams. In so doing, players hacked the original play mode and consequently impacted on the gameplay.

Referring to studies on Massively Multiplayer Online Games—MMOGs—and Massively Multiplayer Online Role-Playing Games—MMORPGs—we could define the condition of a single player in LBMGs as that of being *alone-together* (Ducheneaut et al. 2006). Players roam the virtual or physical space alone and act following their in-game goals, but in the meantime, they are aware of what other players are doing or have done and how these actions impact on the game. Players are actually alone, and play alone, but feel surrounded by other players and part of a social group: in other words, they can feel the social presence of the others, also without interacting directly with them. The sense of social presence is also at the basis of the pioneer of LBMGs, the *Geocaching* game, which employs the logbook with a twofold aim: as a reward for the player, who can leave a trace of her achievement, but also to show the track of all the players who found the spot. The feeling of being surrounded by other players, despite playing alone, is also common to check-in games such as *Foursquare*: to become the *major* of a place, players must check-in there more than other players.

As already stated, the students of the Augmented Reality and Mobile Experience course were asked to design LBMGs to be played in group; thus, no inferences can be drawn by the analysis of the 44 LBMGs nor examples can be provided regarding the *alone* social configuration.

On the contrary, the *together* mode was frequently proposed as social configuration in the play experience. This mode was often chosen because it facilitates the creation of new social ties among players, or the reinforcement of existing ones by triggering and keeping alive the sense of the group during the entire play experience. The team of players is required to stay together for the whole duration of the game, facing the same challenges and sharing the play experience.

Often, the *together* mode is a key to the very meaning of the game, since it leads to the exact group dynamics at which designers want to point the players' attention. The game *The Rapture* (Conti et al. 2015), for example, aims at sensitizing players on the in-group dynamics that characterize social entities such as the *black bloc*, which frequently uses violence as a form of protest. In the game, the unaware players are asked to help each other in order to destroy objects hiding codes that must be inserted in the mobile device in order to proceed with the game. Players are convinced to act for the best, since they are told to be heroes in a post-apocalyptical world, who fight against a totalitarian regime. In doing so, they continuously

Fig. 4.2   Team of players of *The Rapture* (Conti et al. 2015)

encourage each other, enforcing the sense of belonging and the idea of group (Fig. 4.2). Despite the final *plot twist*, which changes the players' point of view leading them to reinterpret what they did under another light, the team unwittingly spend more than one hour in building social relationships and enforcing the sense of being a group.

Analogously, the game *Beta Sigma PI* (Carbone et al. 2015) takes inspiration from American fraternity and sorority initiation rituals and aims at creating awareness about the issue of indirect bullying (Fig. 4.3). Wearing unwittingly the clothes of bullies, players are involved into an escalation of "innocent" actions which symbolize common conditions of persecution. The sense of being part of a restricted circle, sharing the same ideas and conforming to codified rules, even if unfair, is enforced by the game dynamics: players are asked to perform actions that make the other members of the group laugh at the expenses of persons that are out of the social group of players, passers-by not involved in the game. The final reward is being accepted as a member of the fraternity, which, at the end, players discover to be nothing more than a group of bullies.

Fig. 4.3   Players of *Beta Sigma PI* (Carbone et al. 2015)

Beta Sigma PI opens to non-players, thus entering the group of pervasive LBMGs. Nevertheless, we include it in this section since we want to focus on how the group of player configures socially to face the challenges that the game proposes, relaying upon each other's strengths, and looking for the approval of the group. Informal interviews of players who tested the above cited games show that players experienced the sense of being part of a circle, to be on the same boat—echoing the Italian motto "we are all in this together"—harnessing the sense of immersion in the fictional world and the engagement produced by the game mechanics.

Huizinga's definition of play, as something that promotes the formation of a social group, surrounded by secrecy and keen to mark its difference from the common world (Huizinga 1938), in these games is caught at its very essence, and consciously exploited by the designers in order to convey meaning, namely to raise players' awareness on certain insane group dynamics.

The *together* social configuration is the starting point for the *not together* mode, which is based upon a dynamic movement back and forth between the condition of being all together, and that of separation of some or all the components of the group. Splitting the group for a quest or for completing a task is very common in MMORPGs (Ducheneaut and Moore 2004; Ducheneaut et al. 2006), and similar social dynamics can be proposed in LBMGs by alternating moments of separation with others of reunion. The separation can be favoured by the designer and well suits with role-playing LBMGs, since the different abilities linked to each role can justify a temporary separation of one or more players from the group. An example comes from the game *The Origins of Forging* (Belloni et al. 2016) that brings players back to the times of Greek Gods in a discovery of the mythological roots of craftsmanship. The aim of the game is to sensitize players on the progressive disappearing of craftsmanship, asking them to play the role of ancient Gods. By wearing the clothes of Artemis, Ares, Poseidon and Hermes, the players should collect items to be given to Hephaestus in order to have their iconic object forged: the sword for Ares, the bow for Artemis, the trident for Poseidon and the caduceus for Hermes. Each player is asked to face challenges that only his or her "divine power" allow to overcome, and in doing so, they temporarily separate from the group (Fig. 4.4). The separation is therefore functional to the gameplay and mainly aimed at giving each player his or her moment of glory and to assure the active participation of every player in the group to the game experience.

Other times, as described for the *together* configuration, the *not together* mode can also become a way to convey the message of the game, embedding part of the meaning in the act of splitting the group. The game *The Divine Tragedy* (Fornaro et al. 2015) deals with the issue of stalking with an ironic attitude, making the players wear the roles, and clothes, of Dante and Beatrice. Two players act as a young Dante who lives his student life in the campus, while other two players impersonate Beatrice who cannot bear the end of their relationship and continuously follows Dante, in a sequence of separations and encounters. The escalation of apparently casual encounters is a key to convey the meaning of the game. The players wearing the role of Dante can feel the increasing pressure that commonly characterizes the victims of stalking, while those impersonating Beatrice are

Fig. 4.4 Team of players of *The Origins of Forging* (Belloni et al. 2016) deciding how to proceed with the game (left); a player wearing the role of Hermes running to complete her personal task (right)

convinced to act correctly. The social configuration of players, crucial to the gameplay, becomes here relevant for the interpretation of the real meaning of the game. Each new encounter, after a separation, acquires a specific meaning in the gameplay, since it marks the increasing pressure of the stalker on the victim.

Sometimes, the separation acts together as a metaphor and as a physical condition of exclusion. In the games *The Treasures of Captain Torment* (Boni et al. 2015) and *The Lost Papyrus* (Benedetti et al. 2015), the separation is the representation of depression and illness, respectively. In these cases, players feel the separation as a deprivation and a marginalization from the group, and through the span of the game, it leads them to personally experience the sense of being set apart.

Although the *not together* mode fosters the separation of players in the gameplay, the sense of being part of a group is still perceived, as the informal interviews with players show. The acts of separation and reunion of the group, indeed, are pointed by the players as powerful ways to increase the feeling of being part of a team.

Granovetter describes the strength of a social tie as an interrelated combination of the amount of time, the emotional intensity and the intimacy which characterize the tie (Granovetter 1973). Therefore, the *together* and *not together* social configurations are ways to enhance the strength of social ties, since players spend together hours playing a game with a high level of emotional involvement and often experiencing intimacy with other players. A confidence elicited by the game mechanics or sometimes forced by the designers, who can ask players to go a step further from the common in-person social relation between players, and get engaged in a direct physical contact that challenges implicit social norms. In the game *Keep (the Date) Safe* (Aufiero et al. 2015), aimed at sensitizing players on the pros and cons of the most common contraceptive systems, players are frequently involved in activities and mini-games that ask for transgressive social behaviours (Fig. 4.5). This pushes players to go far beyond their comfort zone, invading that space that proxemics (Hall 1990) would define as personal.

Fig. 4.5 Players of *Keep (the Date) Safe* (Aufiero et al. 2015*)* following the game instructions on the mobile device (left); twister game reinterpreted to force physical social interaction between players (right)

Being a group means also discussing together about the game when it is over, making meaning out of the experience just done. Informal interviews as well as our participant observation made evident that players, especially when involved in persuasive LBMGs, comprehend completely the meaning of the game only when discussing together and sharing their own point of view. Discussing messages, perspectives, experiences is a crucial moment of disclosure that can be enlightening. Stimulated by confrontation between those who shared an experience, players can discover multiple layers of interpretation. This condition is particularly evident when LBMGs are designed following the stealth approach proposed by Kaufman and Flanagan (2015; Sect. 6.2.2), and the end of the game becomes an epiphany, a revelation of the real meaning embedded in the game, and intermixed or obfuscated in the game experience. The unexpected revelation works as a key (Sect. 6.2.3) to reinterpret the experience but also as a trigger for social discussion and confrontation.

The last mode of social configuration, *not alone*, is by far the less represented since no commercial LBMGs are purposely designed to engage single players with other players, breaking the anonymity that characterizes them. Games such as *Pokémon Go* can favour spontaneous social aggregation around points of interest, places where the Pokémons are, but do not prompt direct social engagement between players by gameplay. Nevertheless, the *not alone* mode could be easily integrated in commercial LBMGs by exploiting the potential of mobile devices of being location and context aware. Concurring with Montola and colleagues, the interaction with other players can transform the game into something bigger than an on-screen experience (Montola et al. 2009), and working on the social configuration of players can be a way to achieve this aim. Of course, privacy and safety issues should be carefully considered when compelling players, by gameplay, to interact in person with other unknown players across the city.

4.2 Outside the Magic Circle: Interacting with Strangers

Players' interaction with non-players is a relevant issue in LBMGs and opens to the broad topic of pervasive games that expand the magic circle by including outsiders in the gameplay. As a matter of fact, LBMGs, and in particular urban LBMGs, can be triggers of social engagement with non-players, bystanders and passers-by not (formally) taking part to the game.

Breaking the borders of the magic circle to interact with persons out of the restricted circle of players could mean coming in contact with strangers, outsiders and newcomers that are neither friends nor acquaintances, but persons with whom players have no ties.

George Simmel describes the social figure of the stranger, as a unity of nearness and remoteness. The stranger is close to us, since we share with him or her common features, but the stranger is also far from us, because those features are also shared by a great amount of people (Simmel 1950). According to the German sociologist, the stranger is part of the group, but at the same time, he is outside and acts as a means of confrontation. Being a stranger is therefore a positive relation and a specific form of interaction that can be exploited for the gameplay. Indeed, to change the status of a stranger and transform her or him into an acquaintance, we should break the anonymity through a conversation (Simmel 1950). Dialogue is therefore the desirable result of game mechanics that aim at breaking the anonymity of players and trigger social engagement with non-players, the strangers and vice versa.

The LBMGs designed by the students, and here objects of study, were all tested and played within the university campus, so describing non-players as complete strangers could be misleading. In this sense, the definition of familiar stranger proposed by Milgram (1977) could be fair. The familiar stranger is someone we meet every morning at the bus stop or at the campus, but we have never talked with. We do not know him or her, but we notice if he or she is not there. As it happens with the social figure of the stranger, the dialogue is the means for breaking the anonymity with the familiar strangers.

Therefore, LBMGs can play a relevant role in favouring direct social engagement with strangers and familiar strangers, and in doing so, enlarging the network of acquaintances and establishing that kind of social ties that Granovetter (1973) calls weak, in opposition to the strong ones characterizing friends and relatives.

Capitalizing on Granovetter's theory of the strength of weak ties and on Rogers's work on the diffusion of innovations (Rogers 1962), Kleijnen et al. (2009) claim that people with just strong ties can access information only from a small group, while people with weak ties may enjoy more contacts and are more likely to obtain diverse information. Furthermore, pre-established group of players, as it commonly happens for event LBMGs, is likely to be constituted by similar people and relies on what Lazarsfeld and Merton call *homophily* (Lazarsfeld and Merton 1954). Homophily is intended by the authors as similarity between persons, acting as an important facilitator of relational ties since it eases communication (Kleijnen et al. 2009).

On the contrary, the social engagement with strangers could result in the building of what Everett Rogers calls *heterophilous networks* that, according to the author, are effective in spreading innovation (Rogers 1962).

LBMGs, and their mechanics in particular, can act as personal entry points for engaging confidently with strangers and consequently scaffolding social experiences. As a matter of fact, mixing game mechanics aimed at enhancing the sense of being part of a group (strong ties) with others, which trigger engagement with outsiders (weak ties), seem to be a good way of balancing the two natures of LBMGs. On the one hand, these games can favour the social bonding of players and, on the other, they can serve as triggers of social engagement with strangers.

The game *The Infection* (Bassanese et al. 2015) is an example where these two souls of LBMGs are successfully combined. As described in Sect. 3.3, it deals with the issue of STDs—sexually transmitted diseases—and involves players in unexpected contacts with non-players. The city is transformed into a fictional world where four bosses—symbolizing four diseases—must infect as many people as possible. By using a smartphone and several in-game physical elements, players wear the role of the bosses and act for spreading the diseases.

While reinforcing the sense of being part of a group, the game proposes two kinds of interaction with outsiders. On the one hand, non-players are unaware of being part of the game since one of the missions for players is sticking adhesive labels representing the viruses to passers-by without being caught. On the other hand, outsiders are physically involved in the game and exhorted to overcome the boundaries of their comfort zone when directly asked to explore the body of players looking for stickers of viruses (Fig. 4.6).

In the above-mentioned game, the involvement of outsiders is part of the game and it is required to proceed with the game. However, encounters with non-players can also be accidental or disguised as such. The play experience in LBMGs usually happens across the city, and players may ignore whether the people they meet are part

Fig. 4.6 In *The Infection* (Bassanese et al. 2015) players are asked to paste stickers on unaware passers-by (left) and to explore the body of other players (right) and outsiders looking for stickers

of the game or not (Reid 2008), and how to favour the inclusion of non-players in the gameplay is a clear matter of design. Consequently, game mechanics should be included in the gameplay to foster the emergence of social engagement with outsiders.

Inspired by the work of Reid (2008), Montola (2011) proposes three categories of successful coincidental experiences that involve non-players in LBMGs: *actual coincidences*, as the name clearly suggests, cannot be controlled by the designers; *calculated coincidences* are partially controlled by the designers, while *fabricated coincidences* are expected, since included in the gameplay. To clarify the category of *calculated coincidences*, Montola refers to *Uncle Roy All Around You* that tells the player to follow a black-haired woman, relying on the possibility that one is always in the area (Montola 2011), while *fabricated coincidences* usually involve informed actors that pretend to be outsiders. This last approach could also result in an increased ambiguity, since players may assume that real outsiders are pretenders (Montola 2011).

In line with this reasoning, Montola maintains that coincidences are unlikely to be generated from the game technology, since mobile phone is greatly controlled: coincidental social encounters with outsiders in LBMGs emerge from the social and physical context (Montola 2011) (Fig. 4.7).

Some of the LBMGs designed by students in our didactic activity tried to overcome this limitation, using mobile devices as a way to stimulate coincidental social experiences. An example is the game *Rewind* (Gubbiani et al. 2014) that deals with the issue of ageing. It exploits the uncertainty of GPS positioning system to make players enter different shops and strike up a conversation with the

Fig. 4.7 Players of *Rewind* (Gubbiani et al. 2014) interacting with a shopkeeper

shopkeepers. Only one them is an informed actor who provides players with the right hints, while the others are actual outsiders. Following the *seamful design approach* (Chalmers and Maccoll 2003; Chalmers et al. 2005; Broll and Benford 2005), a technological flow is integrated in the gameplay and exploited to trigger coincidental conversations.

Interacting with outsiders in LBMGs, especially when the encounter is not only based on dialogue, is not always a way to make the game bigger than just a casual on-screen experience (Montola et al. 2009). The forced interaction between players and non-players can also result in a feeling of uneasiness (Mariani 2016). The game *Beta Sigma PI* (Carbone et al. 2015), discussed above as a good example for the building of strong social ties among the players, proposes as one of its tasks to spray outsiders with a water gun. During the playtests, no players completed the mission because they felt nervous about performing this action. Analogously non-players also can be set in a condition of discomfort: in *The Infection* (Bassanese et al. 2015), the task requiring outsiders to explore players body, looking for stickers, was never completed by male outsiders on male players, while no problems were encountered with other gender combinations. Eventually, it resulted in situations of surprise and amusement that triggered curiosity.

In some cases, the game suggests social interactions that extend beyond the usual ones, and it is usually accepted by players who feel allowed to dare; other times, the kind of interaction pulled is perceived as "unacceptable", arising individual constraints. If sometimes this is an aspect that the game is willing to achieve, other times, it brings to emergent play. A clear example is the one of forced social interactions. This condition, indeed, is largely influenced by the cultural and social context in which the game was designed first and played later.

The feeling of discomfort for players engaged in LBMGs is not uncommon, since they can feel uneasy for several reasons, as reported by Montola (2011) that capitalizing on several authors (Bell et al. 2006; Ballagas et al. 2008; Herbst et al. 2008), lists obvious gestures, role-playing, equipment, sound effects and ridiculous actions as sources of uneasiness for players.

Playing a pervasive game, especially an urban pervasive game, involves a partial loss of the protective scaffold of the magic circle, and players could feel under the critical eyes of non-players, since they are behaving out of the ordinary social frame (Goffman 1974). Nevertheless, the loss is partial, because the weird and ridiculous behaviours, the props and the sound effects, at the end, can be a means to declare "this is play" using Bateson's words (1972).

References

Aufiero V, Boffo V, Regàs A, Teruzzi V (2015) Keep (the date) safe [LBMG]. Politecnico di Milano, School of Design, Milan, Italy

Ballagas R, Kuntze A, Walz S (2008) Gaming tourism: lessons from evaluating REXplorer, a pervasive game for tourists. In: Indulska J, Patterson DJ, Rodden T, Ott M (eds) Pervasive computing. Springer, Berlin, pp 244–261

Bassanese G, Bonfarnuzzo L, Pham C, Redana F (2015) The infection [LBMG]. Politecnico di Milano, School of Design, Milan, Italy

Bateson G (1972) Steps to an ecology of mind: collected essays in anthropology, psychiatry, evolution, and epistemology. University of Chicago Press, Chicago

Bell M, Chalmers M, Barkhuus L et al (2006) Interweaving mobile games with everyday life. In: Proceedings of the SIGCHI conference on human factors in computing systems. ACM, New York, NY, USA, pp 417–426

Belloni E, Bucalossi C, Mazzoleni C, Menini M (2016) The origins of forging [LBMG]. Politecnico di Milano, School of Design, Milan, Italy

Benedetti A, De Marco A, Franco Conesa CM, Piatti J (2015) The lost papyrus [LBMG]. Politecnico di Milano, School of Design, Milan, Italy

Boni A, Frizzi G, Taccola S (2015) The treasures of Captain Torment [LBMG]. Politecnico di Milano, School of Design, Milan, Italy

Broll G, Benford S (2005) Seamful design for location-based mobile games. Entertainment computing—ICEC 2005. Springer, Berlin, Heidelberg, pp 155–166

Carbone L, Dell'oro S, Manessi C, Mariani D (2015) Beta sigma PI [LBMG]. Politecnico di Milano, School of Design, Milan, Italy

Chalmers M, Maccoll I (2003) Seamful and seamless design in ubiquitous computing. In: Proceedings of workshop at the crossroads: the interaction of HCI and systems issues in ubicomp 2003

Chalmers M, Bell M, Brown B et al (2005) Gaming on the edge: using seams in ubicomp games. In: Proceedings of the 2005 ACM SIGCHI international conference on advances in computer entertainment technology. ACM, New York, NY, USA, pp 306–309

Conti N, Saracino G, Serbanescu A, Valente N (2015) The rapture [LBMG]. Politecnico di Milano, School of Design, Milan, Italy

Debenedetti S (2003) Investigating the role of companions in the art museum experience. Int J Arts Manag 5:52–63

Ducheneaut N, Moore RJ (2004) The social side of gaming: a study of the interaction patterns in a massively multiplayer online game. CSCW 2004. ACM, New York, pp 360–369

Ducheneaut N, Yee N, Nickell E, Moore RJ (2006) "Alone together?" Exploring the social dynamics of massively multiplayer online games. In: CHI '06 Proceedings of the SIGHCHI conference on human factors in computing systems. ACM, New York, pp 1–10

Fornaro S, Lancini V, Spalenza A (2015) The divine tragedy [LBMG]. Politecnico di Milano, School of Design, Milan, Italy

Goffman E (1974) Frame analysis: an essay on the organization of experience. Harvard University Press, Cambridge, MA

Granovetter M (1973) The strength of weak ties. Am J Sociol 78(6):1360–1380

Gubbiani G, Rico Sanchez-Mateos M, Ronchi E, Rosti A, Tabasso (2014) Rewind [LBMG]. Politecnico di Milano, School of Design, Milan, Italy

Hall ET (1990) The hidden dimension, 27th edn. Anchor, New York

Herbst I, Braun A-K, McCall R, Broll W (2008) TimeWarp: interactive time travel with a mobile mixed reality game. In: Proceedings of the 10th international conference on human computer interaction with mobile devices and services. ACM, New York, NY, USA, pp 235–244

Huizinga J (1938) Homo Ludens, 2002 edition. Giulio Einaudi Editore, Torino, Italy

Kaufman G, Flanagan M (2015) A psychologically "embedded" approach to designing games for prosocial causes. Cyberpsychol J Psychosoc Res Cyberspace 9

Kleijnen M, Lievens A, de Ruyter K, Wetzsel M (2009) Knowledge creation through mobile social networks and its impact on intentions to use innovative mobile services. J Serv Res 12(1): 15–35

Lazarsfeld P, Merton R (1954) Friendship as a social process: a substantive and methodological analysis. In: Berger M, Abel T, Page CH (eds) Freedom and control in modern society. Van Nostrand, New York, pp 18–66

Mariani I (2016) Meaningful negative experiences within games for social change. Designing and analysing games as persuasive communication systems. Dissertation, Politecnico di Milano

Milgram S (1977) The familiar stranger: An aspect of urban anonymity. In: Milgram S (ed) The individual in a social world: Essays and experiments. Addison-Wesley, pp 51–53

Montola M (2011) A ludological view on the pervasive mixed-reality game research paradigm. Pers Ubiquitous Comput 15:3–12. https://doi.org/10.1007/s00779-010-0307-7

Montola M, Stenros J, Waern A (2009) Pervasive games. Experiences on the boundary between life and play. Morgan Kaufmann Publishers, Burlington, MA

Reid J (2008) Design for Coincidence: incorporating real world artifacts in location based games. In: Proceedings of the 3rd international conference on digital interactive media in entertainment and arts. ACM, New York, NY, pp 18–25

Rogers EM (1962) Diffusion of innovations. Free Press, New York, NY

Sicart M (2017) Reality has always been augmented: play and the promises of Pokémon GO. Mob Media Commun 5:30–33. https://doi.org/10.1177/2050157916677863

Simmel G (1950) The stranger. In: Wolff K (ed) The sociology of Georg Simmel. Free Press, New York, NY

Chapter 5
LBMG as Persuasive Medium

Abstract This chapter looks at games as means of communication, and as contexts of meanings, namely spaces where designers can codify, represent and perform meanings that players are asked to interpret and decode, making sense of the game and its layers of sense.

Keywords Persuasive Games · Systems of Communication · Embedded Meanings · Interpretation · Situated Experiences

5.1 Between Play and Ordinary Life: Making Sense Through Games

A growing body of research investigates "games with a purpose" capitalizing on how games can shape perspectives and deal with dilemmas (Sicart 2010a). Games have been studied as potential means to embed values (Flanagan and Nissenbaum 2014) and disseminate authored positions (Bogost 2007), as well as means to influence user and shape their behaviour in directions determined by who crafted the game as a persuasive tool operating conditioning (Fogg 2002). Acknowledging these two diverse approaches to persuasive games, according to the purpose of this book, we focus on the first perspective, looking at LBMGs as ways to explore topics and even challenge existing attitudes.

To start investigating LBMG as an expressive medium able to *persuade* its players, we need to recall that, by nature, games convey meanings. Briefly echoing Huizinga (1938), games and plays are deeply intertwined with human society and the creation of meaning. They act as contexts of representation (Frasca 2003; Salen and Zimmerman 2004) wherein meanings are embedded (Flanagan and Nissenbaum 2014). On this regard, the issue goes forward. On the one hand, designers can codify, construct and perform meanings; on the other hand, they are seized, translated and grasped by players via subjective interpretation (Sicart 2011). According to Winnicott (1971) and Piaget (1962), we play in an attempt to understand the world, what surrounds individuals—socially, culturally and

D. Spallazzo and I. Mariani, *Location-Based Mobile Games*,
PoliMI SpringerBriefs, https://doi.org/10.1007/978-3-319-75256-3_5

Fig. 5.1 Games are artefacts that induce players to interact with specific representations; in so doing they encourage players to grasp meanings and form judgments about the represented system

critically speaking. Playing a game means experiencing (1) what the game represents and (2) what the game is a representation for. As a matter of fact, inducing players to interact with specific representations, games encourage to initiate expressive appropriation of meaning based on forming opinions and even judgments about the very systems represented (Fig. 5.1).

Certain games are to a great extent based on this point and are purposely designed to influence players by mounting arguments in a persuasive way (Bogost 2007). Hence, articulating the reasoning about the different actors that are involved in the process, games emerged for what they are and can be: complex, dynamic systems of communication wherein players make sense and meaning of things. They prove to be systems of particular interest because of their twofold structure, which contains ends and means in the meanwhile, in the shape of the game aims and its rules (Parlett 1999). The point is that those who designs games meant to question matters of social interest actually designs systems able to convey meanings that can open political debate, and even challenge ethical reasoning about the topic addressed. In particular, this reasoning increases pertinence when related to those games that open political questions and challenge ethical reasoning about wicked problems (Rittel and Webber 1973, 1974; Schrier and Gibson 2010; Sicart 2010a). As a matter of fact, by doing that, the game assumes the shape of an ambiguous, "safer" space of exploration, wherein it is possible to make meaningful experiences.

To explain the communication power of the medium, it is necessary to briefly recall the concept of *magic circle* as an intensely debated term in-game studies, between defence and criticism. The term has quickly spread because it intuitively describes the separation between play and non-play; a division between play and ordinary life that is described as totally existent (Malaby 2007) and as ultimately invalid (Consalvo 2009) in the meanwhile. Zimmerman rehearses the concept he contributed to spread in 2004 with the seminal work he co-authored, *Rules of Play*, declaring that an absolute separation between game and non-game is a misinterpretation; on the opposite, games may be context from which meaning can arise (Zimmerman 2012). It is a useful metaphor, a shorthand that represents a more complex set of social relations (Stenros 2012), since it allows players to experience situations and roles even far from their ordinary, with meaningful possibilities in terms of understanding, and important social and cultural consequences (Malaby 2007; Stenros 2012).

Acknowledging that reality and games are different entities, it has also been proved by the researches of scholars as Huizinga (1938), Caillois (1958) and

Goffman (1967) that they are absolutely connected and intertwined. Goffman (1961, 1967), for instance, looks at games as instrumental systems of analysis and observation. He describes them as simplified representations of real-life situations that remove the unnecessary to unveil their real and bare structures. In consequence, the concept of magic circle is presented as an *interaction membrane*, permeable to the world, since reality and games are connected and they run "mutual interferences". Banally, the experiential knowledge we collected over the years is an element can enter the circle and affect the meanings of the game.

Significant examples that concretize this reasoning are those games that ask, for instance, players to negotiate their positions putting themselves in the shoes of perpetrators, with interesting moral and ethical implications. Hence, playing these games allows players to distance from the real life, entering a space wherein they can interact with a representation of the world (or better, parts of it), in a moral or immoral way. Through such exploration of perspectives that differ from their usual, players can gain knowledge that can persist after the game ends, contributing to the creation of opinions as well as judgements.

The peculiar nature of these games asks for a discussion with a twofold perspective (see Sect. 5.2): the one of (1) the designer who structures the game as a means of communication, and the one of (2) the subject who plays the game and receives its message(s). Advocating for play(er) experiences that encourage commentaries and invite confrontations, the game design activity is grounded on activating processes of reflexion that invite the player to make sense in and out the game (see Sect. 5.3). Their intent goes beyond the communication of meanings to an audience. Being meant for more than play for entertainment, they connect the player to the everyday life and practices, providing opportunities for questioning former ideas, thinking and discuss moral or ethical issues. Recalling Geertz's (1973), the game acts as a *social metacomment*. As a representation of something—that can be a situation, a system or a process—it goes forward its mere depiction. It is intended to bring to light and *express* issues that are not evident, or even usually inaccessible because "sealed" to public observation and reasoning.

5.2 Games as Communication Systems

The practice of sense-making (Weick 1995) requires players to grasp the message the game designer embedded in the game and figure out its meaning. A process that is more complex than it appears (Fig. 5.2).

Because of their intent to convey meaning, we can observe and investigate several games as communication systems. Recalling Shannon–Weaver's model of communication (Shannon and Weaver 1949), communication consists of a sender who transfers a message to a receiver through an appropriate transmission channel; a message that the sender codes and the receiver has to decode. Because of the aim of persuasive games (Bogost 2007), and because of the nature of the game as an artefact (Salen and Zimmerman 2004), we can look at games as sort of

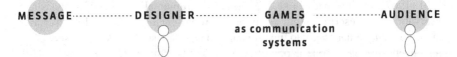

Fig. 5.2 Actors and elements that interact when games act as communication systems

conversations that happen between who designs—as the sender of a message/
meaning—and who plays—as the designated receiver. However, the message/
meaning is not simply "told", but it is embedded in a larger and more complex
system that is the game itself, and it emerges from the play experience, as a result of
the interpretation of the player. Concurring with Galbiati (2005), the Shannon–
Weaver model of communication results not totally appropriate to represent social
processes because unable to properly take into account semantic, syntactic and
pragmatic aspects that contribute to the generation of not-always-predictable
feedback. By focusing their attention on the technological aspects of the process,
the authors Shannon and Weaver (1949) succeeded in outlining a rigorous and
linear integrated model, but have limits concerning the communication vetted from
the perspective of its actors.

Analysing the process through a communication-design-oriented perspective,
Galbiati (2005) points out that an important element of difference is the way the
message is "received"—which actually means "interpreted". Interpretation is in turn
a process based on making sense relying on cultural, social and personal (among the
others) knowledge. In this sense, the communication is imbued of subjective and
contextual variables that results in diverse possibilities to receive the same message,
according to whom is reading it. A perspective investigated by the linguist and
literary theorist Jakobson (1961) who identifies six functions of language—also
named communication functions—that make an act of verbal communication
effective:

- *referential*, connected contextual information,
- *aesthetic/poetic*, linked to the message as auto-reflection,
- *emotive*, associated to the sender' self-expression,
- *conative*, associated to the receiver that takes part to the communication by
  reacting,
- *phatic*, connected to the channel that transfers the message,
- *metalingual*, linked to the code shared between sender and receiver that allows
  the production of meaning.

Observing the game as a means to convey meanings, it is evident the parallelism
between the structure of our verbal communication and how such games work
(Mariani 2016). Hence, merging the reasoning on Shannon–Weaver's and
Jakobson's communication models and adapting it to games, we can associate to

each of these functions an element that characterizes how games work as systems that communicate conveying meanings.

Figure 5.3 represents how the *game designer* (1) aims to transfer (2) *a message* to *the player* (4) who *decodes* (6) it. In the model, the game is used as a (5) *channel*, and its rules and mechanics are the elements through which meanings are embedded and conveyed. Namely, they are in charge of the (3) *transmission* of the message. Then, this entire process happens within a socio-cultural (7) *context* that influences the way the player reads the game and interprets its elements and messages. This is due to the fact that the player, as an individual, negotiates the meanings embedded in the game relying on his or her personal and cultural background (see Sect. 4.2).

Playing, the player activates a process of both reception and interpretation of the game itself, and of the message it embeds, playing a fundamental function in the production of meaning. This production of sense/meaning-making that, in short, depends on the subject's attribution of meaning that in turn is determined by socio-cultural and personal factors, as ethics, moral and background knowledge. The premise is that as individuals we create sense starting a sort of negotiation of meanings between the game, its elements and the real. Meanings are attributed as a consequence of an act of decodification, where players, accordingly to their

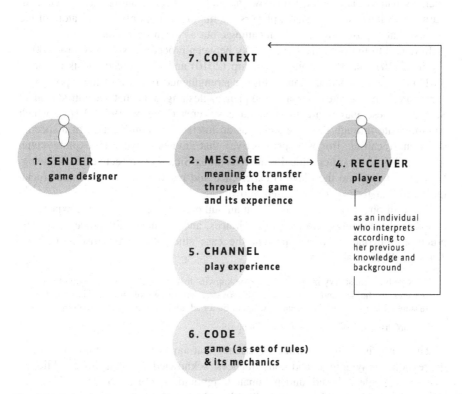

Fig. 5.3 Game elements within Jakobson's and Shannon–Weaver's communication models

personal and cultural background, run interpretations that result into understanding that can stay within the game or can extend to the real world and its practices (Lavender 2008). Once again, the central role of subjectivity (the player) in grasping sense out of games as complex systems with a frequent political and ethical relevance requires to critically discuss the role of proceduralist rhetoric and its limits (Sicart 2011; Mariani 2016).

The argument about games' procedural nature and players' function in reconstructing the meaning embedded in games by means of their rules has been questioned and problematized over the years. However, a specific voice stands out from the choir. It is the game designer and researcher Miguel Sicart who wrote the well-known article *Against Procedurality* (2011) opening the debate about the validity of the proceduralism as a *procedure*. Proceduralism in games relies on the assumption that players understand arguments embedded in the game rules, and then expressed and communicated through the gameplay (Bogost 2007). Sicart outlines the limits of this rhetoric when it is applied to the design and analysis of ethics and politics in games, nurturing a perspective that is inclusive of the interpreter. He asks to look at games as systems where (1) the presence of the player, and (2) play as fundamental elements of the political and ethical relevance of games, take a central position. They do affect messages, because of who is the player, and the way she interacts with a game impact on how the game and its meanings are understood, learned and internalized. Hence, players are important not only as activators of the process that exposes the in-game meanings, but as meaning-makers.

In regard to the considerable diversity between proceduralists as Bogost (2007), Swain (2010), and Brathwaite and Sharp (2010) and anti-proceduralists as Sicart (2011), we take a median stance, further strengthened because of the typology of games here investigated. As a consequence, fleshing out what we introduced in Sect. 5.1, these games are metalanguage systems (Bateson 1955, 1972), which communicate through their gameplay as an interactive simplification of a specific situation (recalling Bogost's perspective) that invites players to explore civic, social, political, moral or ethical issues (recalling Sicart's perspective). In so doing, both mechanics, and the play(er) experience and personal background are crucial means to transfer experience-based knowledge.

To go through what it means design an authored proceduralist game experience that embeds meanings into its design choices, and to unpack the process of interpretations beyond them, we present the case study *The Treasures of Captain Torment* (Boni et al. 2015).

> What ye be doin, scurvy pirate? Don't ye be knowin' that Captain Torment has hidden his treasures in the sea and now waters are crowded with crews of pirates. Find the lost treasures, but pay attention because rush can be a bad advisor and take ye into error!
>
> Forward troops. Let's board for this adventure!

This game is a situated urban experience that aims to sensitize players towards depression, its symptoms and consequences, acknowledging that this condition is frequently neglected and underestimated, ignored, confused or lessened in its effects. To address this, mental disorder characterized by severe feelings of hopelessness and inadequacy, the game is built as a treasure hunt, and players are asked to face a series of tests and obstacle to get to discover where the treasure is located and possibly its content. To transfer the intended messages, the design choice is based on a reasoning that has been formalized by Kaufman and Flanagan (2015), Kaufman et al. (2015); a reasoning that is appropriately explored in Chap. 6 , but that is here introduced to explain why the game mechanics are based upon the use of quite hilarious and awkward dynamics, aided by the interaction with different accessories (as a real cardboard boat, swords, pirate flags and so on). The game is built on a metaphor that is aimed at distracting players from understanding the real issue covered, and hence activate more or less consciously their biases on the topic (Kaufman et al. 2015). Accordingly, also the tone of the narrative is deliberately colloquial and humorous. The story is told by the ghost of Captain Torment: he introduces players to the imaginary world of pirates and provides them instructions to start the adventure. Set in Politecnico di Milano, Bovisa Campus, the game transforms the university area into a sea with islands, bays, eddies and churning adventures to face. In teams of four, players are put in the shoes of as much pirates who board a boat and start tackling several tests to reach the infamous treasure of Captain Torment. At first, they get equipped with some accessories (sword, mop, pirate hat, treasure maps), and they progressively gain extra elements that further increase player's immersion in the narrative (a 2-m cardboard boat, two hooks and a blindfold). Recalling the communication model proposed in Fig. 5.3, each of these objects as well as certain in-game situations codify features typical of depression that the player should grasp and decodify (6). They are elements of a larger message (2) that the game designer (1) translated into game mechanics and elements (5). Example[1] of such meaningful objects and situations are:

- pirates' hook and eye patch symbolize mutilations and distorted view of reality,
- the Bermuda Triangle alludes to the vortex of depression that prevents the achievement of a goal,
- the Siren song stands for the self-conviction of being not able to think clearly.

The crew braves different tests and obstacles that lead one of the players (the vice-captain) to suffer a climax of events, peaking in the total loss of self-awareness and the inability of finding a way out. At this point the crew is at a crossroads that is a crucial, meaningful point of the game: players have to choose between a shortcut that left someone behind, or a longer path that keeps the crew united but put at risk the possibility to reach the treasure in time. The shorter route is here the way the designers translated abandonment into game mechanics. Simplifying, the sacrifice of the vice-captain is used as a lure to distract the sharks that infest the sea around them, allowing the rest of the crew to achieve the treasure within the stipulated time. A tragic act stands for giving up with the ill person, abandoning someone who is perceived as a weight, in order to simplify one's life. On the contrary, the longest path is the acceptation of the depressed person.

---

[1]Further details on objects and narrative elements as metaphors are explained in Sect. 6.2.3.

The game has been playtested several times, focusing on understanding if players effectively comprehended the message embedded in the game in the correct way. Particular attention was indeed posed on understanding how players grasped the extreme contrast between the fun on the ground of the play(er) experience, and the seriousness of the subject addressed. Significant results were gathered especially in respect of the final when the unexpected disclosure asked players to review the steps of the game, re-attributing meanings to the diverse actions taken.

Some of the comments concern indeed how players translated the mechanics, once disclosed the real topic of the game: how they attributed meaning to the various game elements, what message they received, and if it was consistent with the designers' expectations. Example of comments is:

> I've been very moved and struck. Getting immersed in the fictional world was pleasant and the experience was really satisfying. The tests were engaging and very well-structured. Spectacular game material

> The game was fun. The real message became clear just at the end. It created interesting and involving dynamics among the members of the crew: we desired to help each other. Travelling around with the boat and being dressed in pirate style increased the sense of belonging to a group; then the fact of literally wearing a random handicap made the experience even more engaging.

> The tests were engaging, fun and challenging. The narrative was very involving and compelling, it became further fraught with sense after the final disclosure. Things acquired a new sense.

Solicited by the gameplay, players decodified the meanings embedded and started a negotiation of sense that extended outside the game. According to their own sensibility and to their personal and cultural background, they subjectively grasped different layers of meaning. However, we noticed that a deeper comprehension emerged when players started a confrontation about what they individually understood, and how they led back the different mechanics to the symptoms and effects of depression. In so doing, they started a critical discussion that moved beyond the comprehension of the game itself. Especially, the final revelation and the way they faced its crossroad resulted into relevant ethical reasoning. Significant is also the fact that because of the actions they felt free to take when playing; most of the players interviewed often felt the need to justify their actions. They felt the need to momentarily take the distance from the "ethical meaning" of such actions, stressing the fact that it was a game, and that in the real life they would have behaved in a different—and more ethical—way. A decision that depended on the fact that in the game the *real* sense of leaving one member of their crew behind was not so explicit.

According to the experience observed and taking the comments gathered into consideration, the fact of playing being situated in the space, performing social activities and physically embodying the game characters emerges as crucial in conveying the game message. Expanding the reasoning, what we observed is that persuasive LBMGs further the rhetorical capabilities of play, by creating a fertile space of contamination between in-game experience and its out of the game meaning.

5.3   In and Out: Situated Experiences for Contextual Communications

As said, persuasive games aim at raising our awareness or question our knowledge presenting interactive representations (Frasca 2003) that simulate the processes of certain systems or situations (Bogost 2007). The category tends to be mainly composed of digital games; however, for several years we decided to challenge this trend, taking benefit of the power that comes from being embodied in the game. Hence, to better explain the communicative and persuasive power of LBMGs, we draw a parallel between the kind of experience created by (1) digital games, and the one provoked by (2) games that are physically situated—sometimes also linked to the surroundings.

Independently from their high levels of realism, the first are sources of what we can define *indirect experiences*, since they drag players into an experience that is a mediated process. On the contrary, situated games count on physical, *direct experiences*. Experiences *in* the space, and sometime also *of* the space, which require a curious temporary detachment from everyday life, since it persists the awareness of being playing in a game, that however is taking place in the everyday space. On the one side, it means that players are really asked to move in the urban space, with consequences in terms of design: efforts are physical, and the relationship between distances and timings is a crucial element in the gameplay. On the other, this condition can be stressed and used to produce an often underestimated consequence in terms of emotional identification. Even if the game is a medium, and as such it mediates the reality, situated experiences can contribute to activate further layers of meaning that go forward the development of awareness of the surroundings. Recalling the example explained above, players in the shoes of pirates move in the physical space carrying around an unwieldy and cumbersome cardboard boat and wear dress-up clothes and objects that increase their identification and embodiment in the game-roles.

Taking the shape of performances, these games are built on physical embodiment. Players are empowered with a physical agency that activates remarkable processes of *immersion* (Murray 1997; Ryan 2001): they are physical agents responsible of activating the game as they move in the space and interact with *things* (Chap. 6). Such circumstances push forward what Gee defines *projective identities* (Gee 2003). According to this concept, we simultaneously identify ourselves as players (1) with real-world identities, (2) as members of a group with a virtual identity and (3) with in-game characters. These three levels, and the third in particular, are significantly challenged when playing LBMGs, since we are often asked to move in the space wearing the shoes of in-game characters and play certain roles in consequence. And it often comes alongside the request to embrace certain stances and explore alternatives, activating moral and ethical reasoning (Sicart 2010b). Our first-person embodied involvement can invite us to move beyond mere acknowledgement, towards understanding and empathy. What happens during the game as a "momentary detachment from reality" can invite us to embrace new degrees of freedom that allow us to transit across ordinary behavioural categories, and discover different roles (Geertz 1973), as well as cognitive, attitudinal frames (Bogost 2007; Kelly 1955), towards an experience-based knowledge.

On this regard, a case study worth of notice is *The Fellowship of the Umbrella* (Bianchini et al. 2014). It is an undercover LBMG about disabilities, disguised as an epic high-fantasy-based game, where embodiment plays a key role. Set in Middle-earth, the game tells the story of the members of the Fellowship of the Umbrella, an extraordinary magician, a wise dwarf, a powerful beech and a sharp elf, and their adventures to find some keys and reach the treasure. The game starts with a book that introduces players to the story, and these four characters explaining their role in the game, as well as their specific features, contributing to nurturing the *narrative immersion* (Thon 2008) into the story itself, and getting to know the limits and the behaviour rules players are asked to attain to.

In this game, embodiment becomes a way to make players experience visual, motoric, auditive and expressive impairments. This game plays around the term "exceptional" that is conventionally used to make reference to people who are gifted, as well as to people with disabilities. The magician who can not speak because he is "in a water world" is a personification of a dumb; the beech tree, slowed down in its movements because of its rooted knowledge, represents a physical disability; the dwarf is the keeper of several artefacts useful to progress in the game, and it is visually impaired; then the elf, who is able to listen what is inaudible to the others, is deaf (Fig. 5.4).

The disclosure of the real meaning of the roles players embodied happens just at the end of the game, providing a key to re-interpret the entire game, as well as the meaning of the actions performed. As a matter of fact, it is suddenly brought to light the reasons behind players' in-game behaviours and limits, showing such limitations/restrictions as representations of real handicaps. A disclosure that however does not undermine the narrative immersion and its power, on the opposite, it gives a chance to gain a first-hand awareness and further make sense out of the game. On the other hand, further embodiment is due to the spatial immersion that *The Fellowship of the Umbrella* triggers. Besides the presence of diverse game-obstacles, there is a series of minor but meaningful situations that emerge because the experience is situated in the urban space, and that players have to face to proceed in the game. For example, the presence of architectural barriers required

Fig. 5.4 Four characters of *The Fellowship of the Umbrella* (Bianchini et al. 2014) representing disabilities

players to make further efforts to complete some tasks. This impacted on the overall perception of what it means being in the shoes of someone impaired, providing concrete spatial awareness.

In similar LBMGs, the communicative power of persuasive games is further enhanced because merged with the potentiality of physical immersion. As said, rather than being focused on pure digital interactions, situated pervasive games, as LBMGs expand and overlap with the real world, often taking further advantage from the inclusion of diffuse interactions with physical elements. This also means that they involve players into less-mediated processes wherein experiences require to operate acts of sense-making. Concurring with Montola et al. (2009) if the tasks acquire a sense in a wider context, they do not need to be necessarily difficult.

In consequence of this reasoning, it stems a further typology of immersion that characterize LBMGs, we can call *personal immersion*. It is grounded on a social constructivist foundation that sees individuals as agents steadily engaged in making sense of experiences on the basis of their own cultural backgrounds (Creswell 2008). The concept is based on the fact that as much understanding requires an act of interpretation, analogously player's experience depends on the player as an individual with her own ability to connect in-game meanings with previous knowledge. Recognizing the active role of the player in negotiating meanings, it leads back to what has been acquired through experience or education. As such, it depends on the player's ability to enquire the game as a means of communication and translate it to grasp its real meaning, discerning between what is represented and what it is meant to mean, and advancing interpretation and understanding in consequence. The game as a simulation depicts an interactive model that shows how things work (Bogost 2007), and in so doing it acts analogously to what it represents. According to Anolli (2011), this makes each game the result of a mental construction in which distortions, simplifications and variations can occur. Answering to the implicit question "and what if", the game system results to be a more or less accurate reproduction of the logics that rule the reality or some of its parts. To simulate such logics, games are based on maps and models that trace the behaviour of specific objects or processes.

By nature—being a mental construction—these LBMGs require *interpretative conventions*, namely keys to correctly read and use simulations as repositories of meanings. Hence, it is the dynamic relation between the self (player's knowledge), others (the knowledge of who thought and designed the game) and the application of scientific knowledge (technology) that feeds an ontological dialogue that results into the creation and negotiation of meanings. In so doing, both the interpretation of the game as an overall means of communication, and the extension of in-game meanings to real-life realms ground on what Craik (1943, in Anolli 2011) defines an equivalent structure of relationships.

In the light of this reasoning, it stems the threefold communicative potentialities of LBMGs (Fig. 5.5). Beyond being a space of possibility with narrative dimensions (Salen and Zimmerman 2004; Juul 2005) wherein we can engage in artificial conflicts and explore other roles and perspectives, being aware of being in a safer space somehow distanced from the everyday life, the game assumes the shape of a

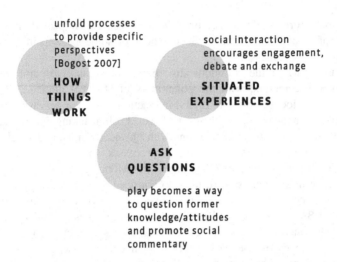

Fig. 5.5 A threefold perspective on what LBMGs as means of communication activate

liminal space. This process consists in staying at the threshold. Without denying the fact that we are still in the realm of reality, we acknowledge the fact of being "playing". This state of ambiguity anthropologically known as *liminality* (Turner 1982) relies on the concept of magic circle introduced above, going beyond a formalist and static idea of play. Persuasive games and games that address social issues move far beyond the old-fashioned simplistic ideas that games are just fun (Lavender 2008), and instead put the player in a paradoxical situation wherein the game becomes a way to understand the reality and reflect on its situations, issues and even inconsistencies.

The concept of liminality extends beyond the realm of the meaning embedded in the game. This further level of liminality depends on the fact that LBMGs take place in a hybrid space, between real and digital, asking players to repeatedly switch between real-life situations, to in-game ones, as detailed in Chap. 6.

Capitalizing on liminality means taking advantage of the game as a safer space of exploration. When we play we know that since we are taking part in a game, what is going on is not *really real*. This frees our ability to imagine, think out of the usual schemata, experiment different levels of sociality and interpersonal connection (Mariani 2016; Mariani and Gandolfi 2016). The game assumes the shape of a liminal space of meaning where conflicts and transformations happen, acting similarly to the functions of ritual and theatre, as described by Turner (1977, 1982): in the first case it exposes and performs existing *social dramas*; in the later it absorbs and re-elaborates ongoing social conflicts. However, it is framed as a *space of potential openness* since players can self-challenge and put themselves to the test. Being the cost of failing within games perceived as low, players sometime also become more open at an emotional level, resulting capable of wearing shoes very different from their own, trying out ideologies that can even diverge from their own, and open their mind to other perspectives. For seminal authors as Huizinga (1938),

Caillois (1958), Goffman (1967), and Turner (1977, 1982), indeed, the play activity assumes a remarkable deep socio-cultural meaning.

In synthesis, the game gives the player the means to freely and (more) safely make experiences within its protected space. As described in Chap. 2, the ability to experiment, fail and make mistakes without *real consequences* outside the magic circle is a feature of primary importance, especially when we look at games as ways to experience the reality (Olson 2010). A space, however, that partially and meaningfully coincide with the real one, speaking of LBMGs. A fact that creates a further condition of *ambiguity*.

References

Anolli L (2011) La sfida della mente multiculturale. Nuove forme di convivenza. Raffaello Cortina, Milan

Bateson G (1955) A theory of play and fantasy; a report on theoretical aspects of the project of study of the role of the paradoxes of abstraction in communication. Psychiatr Res Rep Am Psychiatr Assoc 2:39–51

Bateson G (1972) Steps to an ecology of mind. Ballantine Books, New York

Bianchini S, Mor L, Princigalli V, Sciannamè M (2014) The fellowship of the umbrella [LBMG]. Politecnico di Milano, School of Design, Milan

Bogost I (2007) Persuasive games: the expressive power of videogames. MIT Press, Cambridge

Boni A, Frizzi G, Taccola S (2015) The treasures of captain torment [LBMG]. Politecnico di Milano, School of Design, Milan

Brathwaite B, Sharp J (2010) The mechanic is the message: a post mortem in progress. In: Schrier K, Gibson D (eds) Ethics and game design: teaching values through play. IGI Global, Hershey, pp 311–329

Caillois R (1958) Les jeux et les hommes: le masque et le vertige. Gallimard, Paris

Consalvo M (2009) There is no magic circle. Games Cult 4:408–417

Creswell JW (2008) Research design: qualitative, quantitative, and mixed methods approaches. Sage, Beverly Hills

Flanagan M, Nissenbaum H (2014) Values at play in digital games, Reprint edition. MIT Press, Cambridge

Fogg BJ (2002) Persuasive technology: using computers to change what we think and do. Ubiquity 2002 (December Issue). Article 5: 89–120. https://doi.org/10.1145/764008.763957

Frasca G (2003) Simulation versus narrative. In: Wolf MJP, Perron B (eds) The video game theory reader. Routledge, New York/London, pp 221–235

Galbiati M (2005) Movie Design. Scenari progettuali per il design della comunicazione audiovisiva e multimediale. Edizioni Polidesign, Milan

Gee JP (2003) What video games have to teach us about learning and literacy. Palgrave Macmillan, New York

Geertz C (1973) The interpretation of cultures: selected essays. Basic books, New York

Goffman E (1961) Encounters: two studies in the sociology of interaction. Bobbs-Merrill Company, Indianapolis

Goffman E (1967) Interaction ritual. Doubleday, New York

Huizinga J (1938) Homo Ludens, 2002 edition. Giulio Einaudi Editore, Torino

Jakobson R (1961) Linguistics and communication theory. Proc Symp Appl Math 12:245–252

Juul J (2005) Half-real: video games between real rules and fictional worlds. MIT Press, Cambridge

Kaufman G, Flanagan M (2015) A psychologically "embedded" approach to designing games for prosocial causes. Cyberpsychology 9(3)

Kaufman G, Flanagan M, Seidman M (2015) Creating stealth game interventions for attitude and behavior change: an "Embedded Design" model. In: Proceedings of the 2015 DiGRA conference 12, Article 102. http://www.digra.org/digital-library/publications/creating-stealth-game-interventions-for-attitude-and-behavior-change-an-embedded-design-model. Accessed 20 Dec 2017

Kelly G (1955) Principles of personal construct psychology. Nor, Norton, New York

Lavender TJ (2008) Homeless: it's no game measuring the effectiveness of a persuasive videogame. In: Anonymous proceedings of the 2nd European conference on games based learning academic conferences Limited, p 261

Malaby TM (2007) Beyond play a new approach to games. Games Cult 2(2):95–113

Mariani I (2016) Meaningful negative experiences within games for social change. Designing and analysing games as persuasive communication systems. Dissertation, Politecnico di Milano

Mariani I, Gandolfi E (2016) Negative Experiences as learning trigger: a play experience empirical research on a game for social change case study. Int J Game-Based Learn 6(3):50–73. https://doi.org/10.4018/IJGBL.2016070104

Montola M, Stenros J, Waern A (2009) Pervasive games: theory and design. Morgan Kaufmann Publishers, Burlington

Murray JH (1997) Hamlet on the holodeck: the future of narrative in cyberspace. MIT Press, Cambridge

Olson CK (2010) Children's motivations for video game play in the context of normal development. Rev Gen Psychol 14(2):180–187. https://doi.org/10.1037/a0018984

Parlett DS (1999) The Oxford history of board games. Oxford University Press, New York

Piaget J (1962) Imitation in childhood. Norton, New York

Rittel HW, Webber MM (1973) Dilemmas in a general theory of planning. Policy Sci 4(2):155–169

Rittel H, Webber M (1974) Wicked problems. Man-made Futures 26(1):272–280

Ryan M (2001) Narrative as virtual reality: immersion and interactivity in literature and electronic media. Johns Hopkins University Press, Baltimore

Salen K, Zimmerman E (2004) Rules of play: game design fundamentals. MIT Press, Cambridge

Schrier K, Gibson D (eds) (2010) Ethics and game design: teaching values through play: teaching values through play. IGI Global, Hershey

Shannon C, Weaver W (1949) The mathematical theory of communication. Urbana. University of Illinois Press, Cambridge

Sicart M (2010a) Wicked games: on the design of ethical gameplay. In: Proceedings of the 1st DESIRE network conference on creativity and innovation in design desire network, pp 101–111

Sicart M (2010b) Values between systems: designing ethical gameplay. In: Schrier K, Gibson D (eds) Ethics and game design: teaching values through play. IGI Global, Hershey, pp 1–15

Sicart M (2011) Against procedurality. Game studies 11. http://gamestudies.org/1103/articles/sicart_ap. Accessed 30 Dec 2017

Stenros J (2014) In defence of a magic circle: the social and mental boundaries of play. Transactions of the Digital Games Research Association, 1(2)

Swain C (2010) The mechanic is the message: how to communicate values. In: Schrier K, Gibson D (eds) Ethics and game design: teaching values through play. IGI Global, Hershey, pp 217–235

Thon J (2008) Immersion revisited: on the value of a contested concept. In: Extending experiences-structure, analysis and design of computer game player experience, pp 29–43

Turner V (1977) Process, system, and symbol: a new anthropological synthesis. Daedalus:61–80

Turner V (1982) From ritual to theatre: the human seriousness of play. Paj Publications, New York

Weick KE (1995) Sensemaking in organizations. Sage, Thousand Oaks

Winnicott DW (1971) Playing and reality. Tavistock Publ., London

Zimmerman E (2012) Jerked around by the magic circle: clearing the air ten years later. http://www.gamasutra.com/view/feature/135063/jerked_around_by_the_magic_circle_.php. Accessed 30 Dec 2017

Chapter 6
Stories, Metaphors and Disclosures: A Narrative Perspective Between Interaction and Agency

Abstract The chapter explores the role of narrative in LBMGs. It discusses how games take the shape of storyteller's point of view that translates information into game experience, taking advantage of metaphors, keyings and stealth approaches as ways to transfer messages and meanings without sacrificing players' enjoyment. Avoiding the unconscious activation of psychological barriers, a conscientious use of narrative is indeed functional to communicative impact.

Keywords Narrative · Interaction · Agency · Immersion · Stories
Stealth Approach · Embedded Meanings · Meaning Making · Fictional World
Narrative-Based Games

Narratives are a complex matter of investigation, deeply rooted in different epistemologies, and they are current object of investigation in multiple fields. They are complex by nature, being them a way for organizing experiences and memories, a way for framing and understanding reality and also a way to trigger our imaginative faculty. However, and more interesting for the specific topic of this book, we consider interesting looking at narratives as ways for triggering creativity and imagination, especially undertaking a design perspective. We can give for granted the fact that narratives play a key role in communication, since they evoke and make meanings (triggering meaning-making), and by creating engagement/involvement they are able to convey messages. In terms of design, it is possible to take advantage of the narrative ability to generate knowledge counting on the reader/viewer continuous interpretation of what is consumed; in doing so, narratives go beyond the often straightforward and functional solution-oriented attitude of the design discipline (Venditti 2017; Bertolotti et al. 2016).

It means that designers can and should make use of narratives for opening perspectives on specific, established situations, for depicting future scenarios and for showing existent or potential interactions; in synthesis, for unpacking and communicating complex systems in meaningful and engaging ways. That said, this chapter mostly focuses on games that strongly rely on stories and looks at how to

D. Spallazzo and I. Mariani, *Location-Based Mobile Games*, PoliMI SpringerBriefs, https://doi.org/10.1007/978-3-319-75256-3_6

design narrative-based LBMGs with good stories: neither overwhelming, nor irrelevant, but engaging and immersive.

Narrative is considered as the essence for several human activities, being crucial for construing identities and histories. In consequence, through the past several decades we witnessed an explosion of interest in narrative and stories, making such multifaceted object of inquiry a key matter of investigation for a broad and comprehensive range of research fields. As such it gained the interest of a growing amount of scholars and researchers that delved into the topic from perspectives that range from communication and discourse studies, sociolinguistics, philosophy, literary theory, to media studies, as well as from cognitive and social psychology, sociology, anthropology and ethnography, to artificial intelligence. Among the extended amount of the literature on the topic, we will make particular reference to works of Joseph (1949), Laurel (1991), Ryan (1991, 2006), Ryan and Thon (2014), Murray (1997) and Aarseth (1997), Juul (2005) and Wolf (2012). However, before getting into a deep discourse about games as communication systems that make use of new kinds of storytelling based on stories and metaphors that convey meanings, but dealing with games in general, the main recurrent question uses to be: Are games narratives?

Even if a discourse in terms of narrative theory is not the fulcrum of this book, it is undeniable that narrative serves a major function in some games, being a core component (Frasca 2003; Juul 2005). It is well known that not all games are based on narrative, since not all games rely on stories. However, whereas the medium itself is not narrative-based by definition, stories are notably relevant for specific typologies of games that have stories as their linchpin, as persuasive games, adventure games, live action role-playing games (LARP) and alternate reality games (ARG). By their own nature, such categories quite openly recognize and confirm the central role of narrative. On the opposite, LBMGs as well as urban games and a significant amount of pervasive games have a quite solid tendency to relegate narrative to a design possibility.

6.1 Games and Narrative: A Complex Relationship

When narratives are used in games, they need to be carefully and meaningfully designed, not simply *wrapped around a gameplay* to make if fascinating and attractive (Dena 2017; Dansky 2014). A key feature of narrative is that it exists per se, but cannot be *viewed* independently. A classical point that is covered when dealing with narrative is the fact that by nature stories move across media. According to Chatman (1980), *transposability* means that a story can be translated from one medium to another. It is a key reason for attesting that narrative are structures independent of any medium. As a matter-of-fact, once a story is mastered, it can be spread, preserving its features. Therefore, even if stories are independent units, they are experienced through media, and therefore, to a different extent, linked to them and to the features and affordances of such media.

A mandatory premise is that games and narratives are not very far apart. However, a different matter is how they are tied up. Considering the role of the medium, a first, necessary step to better address the topic of narrative within games requires to unpack their mutual relation. Before digging into how games tell stories, and how designers can craft good stories worth to be experienced, we consider necessary to shortly recall the arguments on the ground of a ten-year-old diatribe that saw as its protagonists the so-called *ludologists* and *narratologists*. Lined up and firmly opposed to each other, the two factions represented two contrasting ways of correlating games and narrative.

For a couple of decades, the game research community witnesses the very existence and lively dispute of two opposed factions. The first considers (video) games as a kind of extension of narrative, taking the name of *narratologist* perspective. The second, known as *ludologist*, totally opposes to the first common assumption, considering games as a different medium, with its own rules and features. Among the plethora of authors who sustained the latter position, Aarseth (1997) stands out questioning that the mere fact of neglecting the presence of differences between games and narratives means ignoring the essential qualities of both categories. Saying that games can include narrative—they feature, for example, narrative introductions and use backstories—does not mean that they *are* narratives.

Sharing some traits with narratives further complicates an already complex situation characterized by often blurred and nebulous boundaries. To unfold the matter, we can echo what argued by Juul in his *Games Telling stories? A brief note on games and narratives* (2001). Looking at games as narratives simply means "framing something as something else". The fact is simple: representing something in a narrative form does not make it narrative (Juul 2001), and in parallel, the fact that games are not narrative does not imply that they cannot include it and use it for some purposes. Similarly, it goes the reasoning—unpacked in the following paragraphs—about games and interactivity: interactivity is a game feature, and it is not what games are.

As both scholars and practitioners, we acknowledge the importance of taking into account the distinguishing properties of the medium, such as its being rule-based, its levels of interactivity with the player and the degree of agency it allows (Laurel 1991). As a matter-of-fact, it is the interactivity that characterizes games that makes them an excellent medium for storytelling (Lebowitz and Klug 2011), differentiating and distancing them from more traditional—or mainstream—media as books and films. Albeit several concepts, tenets and assumptions from the narrative domain can be put into application into the game one, there is a crucial, worth-noticing difference. Juul (2001), the way in which the reader or viewer relates to the storyword is very different from the one in which the player relates to the game world. The nodal point raised by the ludologists is that games and narratives basically provide different ways of interaction. On the topic, Frasca (2003) argued for a distinction between the concepts of representation and simulation, where the first is a key building block of narrative, while the latter is a pivotal and valuable element of digital games. Representation and simulation both rely on several common building blocks (as story, characters, settings, events and so on), but they do have specific tools and practices for conveying meanings and feelings: if representation depicts and outlines a world, simulation models its fundamental and underlying logics. Hence, applying the traditional storytelling model to games appears as inaccurate and restrictive. Games extend indeed the concept of *representation*, being based on different mechanics and on an *alternative semiotical structure* (Frasca 2003). Nevertheless, it is undeniable that games can tell stories (when they are meant to be narrative-based), even through simulation.

In the light of this reasoning, and from a design perspective, in the following it explored the role of stories in narrative-based LBMGs. On the one hand, we will dig into the potentialities and benefits of including narrative in LBMGs, and on the other, we will investigate its boundaries, difficulties and misuses. This is considering that LBMGs simultaneously strongly rely on technology (Montola et al. 2009), interact with the spaces wherein they take place, and hence with their cultural capital.

In the domain of design practice, swinging from game studies to media studies, part of our research was committed to critically enquiring the potentialities of narrative-based LBMGs, and nurturing a constructive discourse around the role of narrative and fictional world in fostering engagement and involvement.

Crafting stories for games is not an easy task, since games are complex systems wherein several aspects have to be stitch together, requiring a set of theoretical underpinnings with perspectives from sociology, anthropology, mathematics and psychology, among the others (Salen and Zimmerman 2004). Then, crafting stories that are mentally and emotionally involving, rather than being merely amusing or engaging, is even more complex. However, to make the situation worse, a further level of difficulty is reached when games are meant to explore selected topics and convey specific messages (Mariani 2016). Messages that players should receive by playing—and hopefully in an accurate way, but this opens a parallel discourse that we will not delve into in this book.

In the following, we unpack the results of our research on the diverse ways of telling stories and engaging players through narrative-based LBMGs that convey specific messages. Therefore, what does it mean to design good stories for LBMGs aimed at transfer specific contents? We cannot really imagine the existence of a straight answer to this question. However, it is possible on the one hand to go through a set of features and guidelines that characterizes games and that should be present also in properly designed narrative-based games, and on the other to identify some practices that allow designers to craft games that embed content and message without sacrificing players' amusement.

Accordingly, we consider key to deal with the concepts of agency, interactivity and immersion in narrative-based games—and LBMGs specifically—while in the second part we will discuss how games take the shape of storyteller's point of view that translates information into game experiences that are forms of staged drama. This is possible by taking advantage of metaphors, keyings and "stealth approaches" (as defined by Kaufman et al. 2015) as ways to transfer messages and meanings without sacrificing players' enjoyment (Fig. 6.1). As we will see, to avoid the unconscious activation of players' psychological barriers, a conscientious use of narrative is indeed functional to the communicative impact of the game.

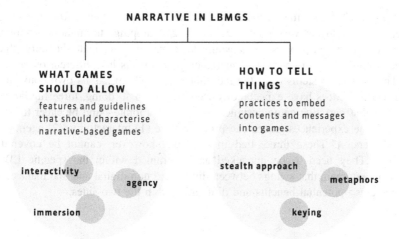

Fig. 6.1 Schema sums up some specific features and guideline to design narrative-based LBMGs, and useful strategies for crafting their stories in a captivating and meaningful fashion

The reasoning that follows stemmed from a specific research conducted on LBMGs, even if some of its points can be extended to narrative-based games in general.

6.2 Enter the Story and Make a Difference

Traditionally, stories are means through which we communicate and share experiences, memories and perspectives (Davenport 2005). They play the same function in narrative-based games, apart from the fact that such games, more than other media, allow players to actively engage in their stories, as if playing temporarily transport them in a world where they can be personally involved in the action, a world where they can take decisions and choose how to proceed.

Designers have at their disposal various possibilities when it comes to design a LBMG story. For example, rehearsing some narrative theory, it can be designed as a linear storyline that the player has to follow, or as a multiple storyline that can be explored and experienced, or again as episodes coming in succession. From a game design perspective, regardless the narrative structure chosen, designing the narrative on the ground of such storylines means providing players with the possibility to try different paths of action to see where they lead, to see whether and to which extent their choices affect the story and its development (Wolf 2012).

At this point, it is crucial to make a distinction between "changing" and "exploring" the story—a difference that often remains unclear to the player itself. Narrative-based games can indeed tell their stories putting the player into the position of (1) impacting on the plot structure to diverse extents, from meaningful alterations, to any or very little ones, or of (2) "jumping" throughout a nonlinear story, meaning that players' choices affect the way they navigate the story (that is hence designed as a hypertext), experiencing the events in a different order.

This reasoning stems from the fact that each medium has its own conventions and that audiences have expectations accordingly. In so doing, three fundamental features that narrative game designers should incorporate—and be eager to obtain through the experiences games can produce—are (1) interactivity, (2) agency and (3) immersion. These three, tied-up concepts, however, cannot be covered in general. They need to be unpacked and re-framed within the specific LBMG domain, a domain that swings between digital and non-digital and that makes these games grasp potential benefits and difficulties from the two sides.

6.2.1 Narrative Across Agency, Interactivity and Immersion

Broadly speaking, and recalling the diatribe summarized above, the logics of the medium make the game narrative differ from the traditional ones. The features that characterize the medium itself impact on its variety of design possibilities. To start digging into the threefold matter that gives the name to this paragraph, and puzzle out its design implications, we must start afresh, and namely from the fact that what links the reader to the story is quite a different matter than what links the player to the game.

Games take some distance from narrative and extend the concept of representation, entailing an active role of the user (Adams 2010; Lebowitz and Klug 2011). In spite of being representations of something, games take the distance from traditional media, being *simulations* (Murray 1997) to which players participate in. Rather than playing a passive role, players are asked *to make the game progress*, taking action. This to say that one of the most evident differences between representation and simulation is the one between observation and participation, between playing a passive role versus an active one, namely what the player is allowed to do with and within the model that the game simulates (Mariani 2016).

A further consequence is that the player is engrossed in a systemic process of interconnected causes and effects that requires the comprehension and internalization of the game's logic to proceed in the gameplay. This further step has an evident, adjunct relevance in terms of meaningfulness once it regards games that address political, civic and social matters. Rehearsing Frasca's reasoning (2001, 2003), by nature games rely on an *alternative semiotical structure*: each game is a system that simulates a model (a piece of a broader system) of which it maintains and replicates some behaviours with evident and distinctive rhetorical possibilities. The key concept of this reasoning is the game ability to *behave*, namely to act and re-act in a specified way in consequence of specific inputs. In other words, simulation is based on answering to stimuli, and it embeds opportunity. It provides the player with the possibility to manipulate the system and observe how it dynamically reacts. In the light of such reasoning, although we recognize the power of representation as a way to understand and explain the world (real or fictional) through narrative, games as simulations are different, since they include responsiveness to choices and actions, and as such good games should be designed. Games are indeed based on the intertwined concepts of *agency* and *interactivity*, where the first is described by Murray (1997) as the ability of computer users to participate in games (as simulations), and the latter is pointed out by Salen and Zimmerman (2004) simply as what defines the game medium itself: the levels of interaction range from the formal one with the game's elements (namely objects and pieces), to the social one between players, to the one with the cultural context in which the play takes place (2003). In a nutshell, it is the interaction of the player that makes the game advance. Then, such a crucial statement relies on the fact that the player is enabled and allowed to take choices that affect how the game proceeds (interacting with the game as a system).

The theoretical rumination advanced so far applies to games in general as well as to LBMGs. However, some specific reasoning specifically contextualized in LBMGs that are narrative-based needs to be unfold and taken into consideration. The very nature of such games makes them in need of specific design and particular attention. Matter-of-factly, LBMGs expand outside traditional gamespaces (de Souza e Silva 2017) and open up to situated and contextual play activities. Hence, their design should harness the features of digital games, the geolocalization technology, as well as some cultural capital of urban games, to transport players into a hybrid world where the real and the digital spheres are overlapped and the boundary between them is fuzzy (Mariani and Spallazzo 2017). In this specific gamespace, having meaningful player's choices means that they ought to be simultaneously significant (Fig. 6.2):

Fig. 6.2 Choices should have a threefold impact. In the meanwhile, they should affect the game, be relevant in terms of narrative, and fitting for the context/space where they are located

1. for the game and its progression (what is doing the player is somehow affecting the game),
2. for its underpinning narrative (the player actions have some kind of impact on the narrative), and
3. for the context in which the game is taking place (at least, to enhance meaningfulness, there should be a reason why that part of the game is set in a specific space/place, rather than another/random one).

Then, focusing on narrative aspects, games can be designed as linear or nonlinear, and they can present multiple endings. The difference between linear and nonlinear lies on the fact that in the first kind of narrative presents the sequence of events that defines the plot following the order of the events as they happen in the story; on the opposite, a nonlinear narrative portrays events in a non-chronological order. This general subdivision works also for LBMGs and results in specific design consequences. Speaking of LBMGs, such narrative typologies are concretely translated in two different ways of structuring (and then playing) a game: in the first case, we have LBMGs designed as a series of locations to visit following a specific order, in the latter as a series of locations with narrative elements that the player can surf and enjoy, following a storyline that emerges accordingly to the places visited.

Taking a gameplay-oriented perspective, such narrative structures, respectively, activate a linear and a nonlinear gameplay. A LBMG with a linear gameplay requires the player to confront with a series of defined challenges. In so doing, every player addresses and overcomes every challenge following the same order. A LBMG with a nonlinear gameplay leads players to face challenges that can be encountered and accomplished in different orders, according to their own choices. It means that different players can encounter the same challenges in a different order. This typology of gameplay recognizes the player a broader freedom, permitting to go across a multiplicity of sequences to complete the game, and hence, different choices and experiences.

Apart from their different structures, linear and nonlinear stories require different design skills and competencies. On the one hand, linear stories are easier to structure and to be made with dramatic consistency and are less expensive in terms of time and development: as said, they consist of one fixed sequence of events. On the other, nonlinear stories are appealing because they provide broader freedom to the player, but result very difficult to be designed bug-free, pleasant and meaningful to play—where meaningful is used both referring to the meanings embedded (Mariani 2016) and the possibility to take choices affecting the gameplay (Salen and Zimmerman 2004). However, in both cases, the game employs one story, with an ending, and the player cannot change or affect the narrative with his/her choices. An interesting compromise are the so-called foldback stories, namely stories split into branches that are then fold back into a single storyline (Fig. 6.3). These stories present some points of convergence, where the different branches merge into inevitable events. In terms of gameplay, such events are narrative point each player, independently from the choice made during the game, is going to encounter (Adams 2010). The strong point of such narrative structure is that it employs

Fig. 6.3 Foldback stories
simplified structure. Adapted
from Adams (2010)

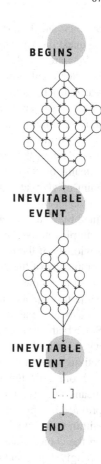

BEGINS

INEVITABLE
EVENT

INEVITABLE
EVENT

[…]

END

nonlinear narrative to give the impression that the gameplay is nonlinear. This result
in a more enduring game experience, as the player can re-play, in order to explore
the different branches that run in parallel between inevitable events.

An example of a foldback story comes from the LBMG *The Divine Tragedy*
(Fornaro et al. 2015; Sect. 4.1), developed in the Augmented Reality and Mobile
Experience course in 2015. To deal with a rather sensitive issue, stalking, the
authors decided to work on a modern, overturned interpretation of the famous
Italian poem by Dante Alighieri, *Divine Comedy*. The narrative develops around an
improbable, bad ending love story between Dante and Beatrice, where the first is
the victim of the undesired attentions of the latter. Beatrice is indeed designed as a
stalker, a rather unusual choice, as victim and stalker roles are generally reversed.
Statistically men are the ones stalking women; however, as thoroughly explained in
Sect. 6.3, challenging ordinary roles, distancing from the topic and hiding layers of
meanings into game activities are design strategies for improving transfer of
meanings. In the light of this reasoning, the game tells the story of Dante,
a twenty-first-century male student who happily lives his day in a university

campus, and Beatrice, a modern female student who does whatever is needed to conquest the guy she has fallen for. This without realizing, at least initially, that her actions are deeply affecting Dante's privacy and way of living.

After conducting a careful desk research and analysis on the topic, students realized that a pivotal element to communicate such a complex, multifaceted and sensitive subject was developing a narrative that employs multiple choices able to activate different storyline. In so doing, in the player perspective, the decisions taken during the gameplay are able to affect the story and the game experience itself. Hence, the game is structured as a role-play game that allows players, paired in teams of two, to play in the different roles of Beatrice or Dante, experiencing a point of view at a time.

Matter-of-factly, the game is structured as a foldback story with a set of prede-fined inevitable events, and different storylines to get there, storylines triggered by multiple choices that players can take in different moments of the gameplay (and obviously of the narrative). What is worthy to note in this case study is the twofold complexity of the design practice: on the one side the one regarding the structure of the story, its inevitable events and its multiple storylines; on the other the fact that the game takes place in the real space, where players playing paired need to simulta-neously move in the gamespace and experience the narrative concurrently (Fig. 6.4). Since both Beatrice and Dante have to be physically present in the spots of the game where and when inevitable events happen, students introduced game elements and activities able to maintain the stories and gameplays of the two players in sync.

From a design perspective, including such interaction when crafting a story means conceiving the narrative as storylines, namely split the story into threads that provide the player with a certain degree of agency within the story, other than within the game. The player can surf a game that presents a narrative that depends on the choices made by playing. Once again, agency and interactivity result to be strongly dependent on the presence of (meaningful) choices. Then, the role of these concepts becomes prominent when dealing with LBMGs that are narrative based.

In each case, none of these narratives preclude the player from performing multiple actions along the way, and *feeling* that she can interact with the story. What matters, according to Adams (2010), is that the player gets the feeling of contributing to the sequence of events that constitutes the story. On the matter, it is remarkably interesting what Juul points out. He states the impossibility to simul-taneously have narration and interactivity, because continuously interactive story is a concept that cannot exist (Juul 2001). The fact is that the very point of narrative-based games is not interaction. The primary function of game narrative is presenting events and situations that cannot be controlled by the player (Adams 2010). Games typically present events and situations that do not allow any agency to the player: they simply need to happen, and the player needs to know them. They are necessary to move the story forward, and the play activity in parallel. Hence, to have games where the narrative goes on while players act (being active, rather than passive observers), any game that includes narrative elements is in need of a proper equilibrium between the desire to act of the player and the need to narrate of the designer (Adams 2010).

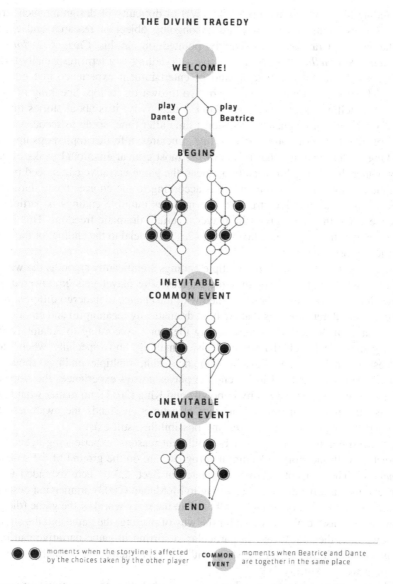

Fig. 6.4 Simplified structure of the foldback story of the LBMG *The Divine Tragedy* (Fornaro et al. 2015)

Then, to conclude the overview, less frequently games can have interactive narratives. When it comes to such games, usually computer games, choices become key to progression to a further level. The player is allowed to go through a storyline that is not predetermined, experiencing a unique story that is based on how she interacted with the story. This means that alternative experiences are activated, responding to the decisions made. The level of difficulty of designing such structures is significant; indeed, they are an ongoing object of research and experimentation. As Crawford very clearly pointed out in his *Chris Crawford on Interactive Storytelling* (2004), interactive storytelling is a term (especially today) often abused since it refers to a kind of entertainment experience that goes far beyond from being a game with a narrative thrown on its top. Speaking of interactive storytelling means speaking of true interactivity: it is about stories that are literally created out of players' decisions, time after time, scene to scene.

In parallel to the discourse on storyline structures, a further topic pops up when speaking of "the ending" that is one of the most critical emotional peaks of each story. Other than being linear and nonlinear, the game narrative can indeed present multiple endings, namely different ends according to the choices made during the game. Framing a narrative structure that includes multiple endings is particularly effective when the game gives the player a certain dramatic freedom. This means that the player has to perceive that her choices are crucial to the ending, or the entire experience loses meanings.

As a matter-of-fact, devising multiple endings significantly smooths the way for an enhanced dramatic role of in-game choices. The player gets empowered with moral choices, especially when provided with the chance to undergo different paths and explore different endings that are both dramatically meaningful and emotionally consistent with the decisions took during the story. According to Adams (2010), when games are largely based on decision-making, and especially when moral choices are included and have a dramatic result, multiple endings should be included. However, even if in the end the player always experiences the story in a linear fashion—as pointed out by Lebowitz and Klug (2011), no matter what line is pursued, the player run through a story at a time—it stands the awareness that further paths, with their adventures and possibilities, still exist.

At two different levels, and for two different reasons, experiencing agency and interacting with the story, affecting its progress, are on the ground of the so-called *immersion*. The concept, firstly investigated in Sect. 5.3, is here extended in reference to its narrative aspects. According to McMahan (2003), immersion describes the player's condition of being "caught up" in the story world of the game (diegetic level), as well as the player's love for the way of playing, the game and the strategy it allows (non-diegetic level). In particular, a stirring in-game narrative can make the player feel in tune with the story with such an intensity to obliterate the surrounding world (Murray 1997).

This reasoning opens up a reflection about what it means experiencing immersion when playing LBMGs. To answer, it becomes necessary to outline a parallel between the experience players undergo when playing digital games, and the one they go through when taking part in games that are physically situated and linked to the surroundings. Murray (1997) describes immersion as a status of *suspension of disbelief* that occurs when the player (user in Murray's original work) participates in simulations that are abstractions and reductions of reality. The reflection on immersion started in the digital game and video game field, in a condition that is "technologically mediated". On the opposite, LBMGs are partly mediated, asking players to move in the real space, supported by mobile device. Hence, in a scenario where digital and video games propose experiences that are mediated and decontextualized, LBMGs offer situated experiences that require the player to move in the urban spaces. The result is a more "direct" experience in and of the spaces. What happens is that the surroundings in which the LBMG takes place usually encounter a perceptual reconfiguration, temporarily acquiring new meanings—the ones the game and/or narrative provide them.

As summed up by Wolf in his *Building Imaginary Worlds* (2012), immersion can be physical, sensual or conceptual. While the first and the second, respectively, refer to the feeling of being physically surrounded by the constructed world, and that what is seen and heard is part of the experience designed, conceptual immersion refers to the vicarious entrance of the user in the imagined, fictional world. As such it largely depends on the user's imagination. Speaking of LBMGs, since they are situated in the real space, by definition, they physically involve players into the game activity, guaranteeing physical immersion. However, what cannot be assumed for granted, but on the opposite is in need of a conscientious activity of design, is the occurrence of conceptual immersion, namely the kind of experience that provides a vivid feeling of going somewhere else, exploring a different place (Wolf 2012). In terms of design, this aspect is as meaningful as complex.

A challenging way to design keeping these three typologies of immersion in mind is conceiving LBMGs including considerable *performative aspects*. Such aspect should be devised not only in order to enhance player's engagement, but to convey meanings, transfer significant elements of the story and increase the sense of *being into the story*, as well as of being playing. Often performative aspects of play become indeed evidences themselves that a game is "on stage" (Bateson 1956, 1972), but in the ordinary, everyday space, and very likely among non-players.

Indeed, even if to a different extent, the interaction of the player with the game should extend to the context, as it extends to the story and the gameplay (Mariani and Spallazzo 2017). In the light of the very nature of LBMGs, especially their being spatially situated, the context has indeed the potentiality to play a recognizable function. A potentiality however often remains a design possibility. Just the fact of playing in a space provokes a shift in the meaning attributed to that space itself that becomes conceptually different for those who play. A matter of interest, often neglected and also stressed by de Souza e Silva (2006), is that there should be a reason for setting a part of the game in a specific space, rather than somewhere else, and this reason should also enhance meaningfulness. The parts of a LBMG should be conceptually linked to the surroundings where they are situated, making significant the fact of happening right there and not somewhere else. This can also be translated into significant interactions with those who are populating spaces. Social interactions can indeed be deliberately and conscientiously designed to foster players to go beyond the common and expected social behaviours. Especially designing such challenging kind of interactions requires a strong and sound narrative: the in-game story and how it is intertwined with the gameplay have indeed to legitimize and justify eventual uncommon way of acting/behaving, making players willing and inclined to do that.

6.3   Disguise It, for Player's Sake!

Much have been written about the combination of narrative and games; on the opposite, the different ways in which games with narrative can transfer knowledge are still a matter in need of exploration. If we can count on seminal authors as Murray (1997), Aarseth (1997), Juul (1999, 2005) and Ryan (2001) for theoretical enquiries on the first topic, when it comes to design practice, a relative lack of references, methodologies and guidelines surfaces.

A branch of our research on LBMGs dug into the role of narrative and fictional worlds in games as elements able to "pull the player into the game" and magnify the game ability to convey contents. To do that, we focused on understanding how narrative and fictional words could facilitate and prompt the comprehension of the meanings embedded into the game.

6.3.1   Say Something, Meaning Something Else: The Case Study of the Treasures of Captain Torment

> Capitan Torment has hidden his treasures in the neighbour and now the "water" around the university are crowded with crews of pirates. Find the lost treasures, but pay attention because rush can be a bad advisor and take you into error!

The Treasures of Captain Torment (Boni et al. 2015) is a LBMG structured as a treasure hunt, where players are challenged to find the legendary infamous treasure of Captain Torment, facing a series of missions masterminded by the Captain himself and meant to test their worth and bravery. Wearing the shoes of pirates, and equipped with swords, hats, flags, a mop and a precious treasure map, teams of four players enter the imaginary world of the Bovisa Sea. Bovisa Sea is a marine world rich of island and bays, vortexes and harbours, sharks and other dangerous presences, geolocated in the yard of the Design School, Bovisa Campus of Politecnico, where digital and real spheres are overlapped.

The game follows the story and instruction that the ghost of Captain Torment is providing via smartphone. The smartphone embodies the ghost who tells the story, guiding players through the environment identified as playground, and through several hints spread in the surroundings. Acting as a storyteller, it leads players through the game and its narrative, helping them to interpret the map and move through the tricks and pitfalls that are set in the gamespace. Hence, as it locates and contextualizes the PoIs to reach, it practically delivers the game narrative elements, contributing to the increase of the players' immersion in the game world (Mariani and Spallazzo 2017; Spallazzo and Mariani 2017). This immersion is continuously sustained by way in which players interact with the game and its narrative, with the elements placed in the environment, and the smartphone itself. The game is indeed provided with a coherent and cohesive narrative (Wolf 2012) that makes a meaningful use of the surroundings and the situated game artefacts as key elements to proceed in the story.

As we will further see in the following, taking full advantage of designing a LBMG, the designers structured a story that conveys meanings and values which not only fit the context in which it is framed, but that is enhanced and magnified by the fact of being situated right there. Since the first design steps, the story is built not to serve the game, providing a mere context or fictional setting, but as a linchpin linked to the meanings to embed, and to the message to convey.

In very truth, *The Treasures of Captain Torment* itself tells a story meaning something else. Going far beyond being a narrative about pirates, it speaks of depression. This LBMG is indeed aimed at increasing awareness of the plight of individuals struggling with such a mental disease whose symptoms are frequently ignored, confused or lessened in their effects and consequences. In particular, it was designed to sensitize players towards the severe feelings of hopelessness, melancholy and inadequacy that characterize this mental condition that often leads depressed persons to be progressively marginalized by relatives, friends and acquaintances.

6.3.2 Stealth Approach: Keep Some Distance and Have a Meaningful Experience

As a matter-of-fact, designing LBMGs that embed values and meanings that are meant to be experienced through play, stories are fundamental. Investigating on the way designers can more intentionally integrate values and meanings into games, the field presents a series of interesting case studies, but they are predominantly video games and digital games. The past decades witnessed the emergence of a plethora of persuasive games and games for social change, meant to address social issues and increase players' awareness of such matters of concern. However, such games, designed to convey serious meaning, and therefore be more than entertainment, revealed to have the tendency to fall into quite common pitfalls that strongly jeopardize their efficacy. Among the others, the main issues that can significantly compromise the effectiveness of such games are their tendency of not being really engaging and pleasant to play and the fact that their topic affects the play desirability. The first is largely due to the fact that the very aim of these games seems to overcome the fact that they should capture and amuse their players. The latter depends on the way we answer to (persuasive) communications: we activate our frames of reference (Goffman 1974); we resist to messages or communications that we perceive as too forthright in their intentions to persuade us (Kaufman et al. 2015); and our bias, especially when dealing with sensitive topics as stereotypes and prejudice (Myers 1987; Deskins 2013).

Among the theoretical ruminations advanced on the topic, we consider particularly remarkable—as well as useful in terms of design practices and methods—the work of Kaufman et al. (2015). Investigating the power of story in crafting powerful experiences able to make a difference in people's attitudes and behaviours, the researchers notice that a literal and bold approach to sensitive issue seems to jeopardize and lower the game capacity to impact on the player. Conducting qualitative and quantitative research on controlled empirical studies, Flanagan and Kaufman point out that employing a direct approach lessens the game ability to transfer messages and/or knowledge. Not posing sufficient psychological distance between the player and the sensitive topic addressed creates expectations and feeds biases, preventing a functional exploration of the issue.

In consequence of this insight and of the following research, Kaufman et al. (2015) identified three strategies to embed meanings and values in the gameplay:

1. *Intermixing* is based on balancing "off-topic" and "on-topic" content to make the embedded message less overt, and more approachable.
2. *Obfuscating* consists of using strategies to divert/focus players' expectations away from the game persuasive intent.
3. *Distancing* employs fiction and metaphor to feed the psychological gap between players and the game persuasive contents; this also means distancing the identities and beliefs of the players from the ones of the game characters.

Acknowledging the potentialities of narrative, and treasuring the results from Flanagan and Kaufman's research, we focused on designing games with compelling and meaningful stories. Stories with multiple layers of meaning—the one of storyline presented and the one of messages embedded—able to raise players' emotions and feelings, activating processes of realizing that trigger reflection and understanding of the topic addressed by the game (Mariani 2016; Mariani and Gandolfi 2016).

The approach becomes clear recalling the game introduced above (Sect. 5.2), *The Treasures of Captain Torment*: while the story told is about pirates, the real content to convey concerns depression, its symptoms and consequences. In order to do that, the game relies on a strong use of metaphors. To build the game, its front story and metaphors, designers researched on the physical and mental features that characterize depression, going through a process of translation and reduction of such disease elements into game elements. In particular, they posed their attention on those minor aspects and situations that meaningfully outline depression, but that tend to be generally little known or underestimated by "normal people".

To transfer the intended messages, the game was designed as a treasure hunt that involves players into a series of missions that entails quite hilarious and awkward dynamics, mainly due to the interaction with different accessories (Fig. 6.5). As said, players start the game equipped with swords, hats, flags, a mop and a precious treasure map. Then, while progressing, they collect further elements meant to increase their immersion in the fictional world, as well as the identification with the character they play (Sect. 5.3). During the game, they end up managing a 2-m long cardboard boat, two hooks and a blindfold.

To effectively convey its meanings, this game employs both distancing and obfuscating strategies. Moreover, the narrative tone of voice and the game elements spread in the gamespace are deliberately designed to be entertaining, and even facetious, aiming at distracting players from understanding the real topic covered. In so doing, players can focus on the game and have an experience that is not affected by players' insidious prejudices and/or biases on the issue stealthy embedded. Such strategies allow players to "consume" a game and its story free of more or less conscious cognitive constraints. Their inclusion facilitates the exploration of dynamics and often even the identification with—or projection in (Gee 2003) unexpected, often distant roles. From the player's perspective, the implicit reassuring feeling of being playing, and hence of being in a protected space, far away from real dangers (Huizinga 1938; Salen and Zimmerman 2004; Juul 2008; Stenros 2014), creates a useful distance and alienation. If the game does not reveal its aim or real topic covered, for the whole span of the game, the player does not realize that the drama staged is actually part of her life. The use of metaphors and fictional worlds enable to increase the distance between the player and what is represented, or the perception of it, as true and real, circumventing players' psychological and cognitive defences. The potentialities and power of such strategy lie on the fact that they can trigger a more receptive mindset for exploring the topic staged, and internalizing the message embedded; topics, situations and mindsets otherwise un-explorable, with remarkable rhetorical potentialities in terms of ethical and moral reflection.

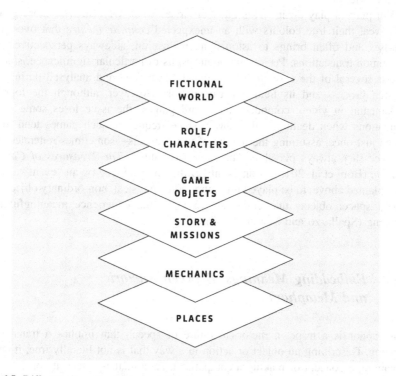

Fig. 6.5 Different elements of the game that can embed metaphors

So far, our discussion agrees with what stated by Kaufman et al. (2015). However, a solid point of their research leans on replayability and enjoyment as stealth approach core mechanics. Although we fully agree on the latter pillar that consists of transferring the message without sacrificing players' enjoyment, we felt the need to take the distance from considering replayability key and crucial a priori.

There are indeed several games that seem to work in the opposite direction than the one suggested by Kaufman et al. (2015). An example is Brenda Brathwaite's series *The Mechanic is the Message*, a set of non-digital games that invites the player to go through some difficult and unexpected topics that take the shape of ethical simulations (Brathwaite and Sharp 2010). The games crafted by the game designer and developer Brenda Romero (previously Brathwaite) rely on a final revelation that hits the player and potentially affect replayability. The fact of employing a final revelation that discloses the real message(s) embedded in the game, as well as the sense beyond the in-game actions (once for all), is certainly impactful and smacking. Surely, such mechanic impedes the player to go again through a similar play experience. Commenting Romero's series, Sharp points out the complexity of replayability, since first-time players were able to enjoy the game and operating under such assumption until they bumped into the meaningful ethical consequences of their actions (Brathwaite and Sharp 2010). Indeed, at a certain point of the game, the real, harsh and cruel implications of their in-game actions become evident.

Will players play again? This question arises when we deal with those games that reveal their true colours with an unexpected *coup de théâtre* that overturns, surprises and often brings to astonishing realization, sideways perspectives and uncommon translations. Per se, the point results of particular significance, since it regards several of the case studies designed, supervised and analysed during this research process and its field experimentation. However, although the topic is fundamental in video, computer and board games, the issue loses some of its foundations when dealing with LBMG, since frequently such games tend to be played just once, assuming the feature of urban events—sometimes reiterated, but not necessary always playable. The same case study, *The Treasures of Captain Torment* (Boni et al. 2015), is an event LBMG characterized by an "event" nature. As explained above, to be played it requires some physical, non-ordinary objects set in real space, objects that contribute to make the experience meaningful and engaging (Spallazzo and Mariani 2017).

6.3.3 Embedding Meanings, Between Rhetoric and Metaphors

A metaphor is a trope, a rhetorical figure of speech that implies a transfer of meaning. Describing an object or action in a way that is not literally true, it helps explaining a concept or making a comparison. As a matter-of-fact, it consists of a replacement of a term with a figurative one, following a symbolic transposition that presents a partial semantic overlap. From the Greek metaphérō, it means "transport".

Over the years, a significant amount of influential scholars investigated how games—as well as other forms of play—can be considered as metaphorical, as representations of something else. From Huizinga (1938), Caillois (1958) and Bateson (1972) who looked at games through an evolutionary perspective, as mirrors of the surrounding culture of which they include some foundational traits, to Goffman (1974) who conceptualized—as expanded below—games as social frameworks and reductions, and to contemporary authors as Bogost (2007), Sicart (2011), Kaufman et al. (2015) who argue that games always refer to something other, further exposing their metaphoric dimension as conceptual extracts, representations and interpretations. Moving forward from this conceptual and high-level rumination about games being metaphorical in general and per se, we intend to draw the attention on the practical, different ways of including meaningful metaphors into games and how they can be rhetorically employed to convey meanings. Designing a game based on metaphors means integrating specific concepts and cognitive meanings into the game itself and asking players to make an effort to grasp, interpret and make sense out of the experience they did, and often out of the choices they made while playing. Therefore, different elements of games can be built as metaphors; and since we are dealing with games that are geolocated, also spaces can be enhanced in their meaning through metaphorical interpretations.

To unpack the topic, we re-analyse a well-known case study of ours, *The Treasures of Captain Torment* (Boni et al. 2015). This LBMG about depression best epitomizes the different degrees of inclusion of metaphors into the design of a game, providing a clear gaze of a practical application. To make ordinary people access an experience that reduces a broader, more complex situation by putting players in new, unexpected roles (Mariani and Spallazzo 2016), the designers of this game decided to largely rely on metaphors. From its mechanics and game objects, to the fictional world, the story itself and its characters, to the places where they are set, this game completely leans on metaphors (Fig. 6.5) that have been identified and structured after a thoughtful phase of desk research and topic analysis.

The fictional world The translation of the topic and its features into the game starts when players access the mobile application (MLA in this case). In the shoes of pirates, a team of four players is welcomed by a mysterious ghost, Captain Torment, into the Bovisa Sea, a piratical fictional world that symbolizes depression itself, characterized by being detached from the rest of the world. This is a world that works following rules and practices that result difficult to be understood by those who look from the outside.

Roles/characters The ghost of Captain Torment represents a sort of inner voice that guides players through their journey. The Captain never physically appears, but is always present when it comes to crucial moments. Then, the players themselves wear the shoes of four diverse characters (Fig. 6.6), with their own specificities, and call for action. When it comes to design meaningful characters or roles, games allow for a plethora of possibilities. However, in this game, the ghost of Captain Torment is the only character fraught with meaning, while the role of each player is significant once linked to the game objects that constitute their equipment.

Game objects According to their role, players start the game equipped with certain game objects that identifies features of the disease and define, during the game, ways of interacting with the surroundings as well as with the others (Spallazzo and Mariani 2017; Mariani and Spallazzo 2017). The swords, the flag and the mop impede, since the beginning, two players to freely use one hand (Fig. 6.6). During the game, according to their role, players end up with a two-metre-long cardboard boat, two hooks and a blindfold. Albeit apparently plain, because of the way of saying, the boat is a tricky metaphor, positive is seen as the representation of the encouraging condition of being all together, sharing difficulties and helping each other; quite neutral if seen as a "we are in this together". The hooks are stand for painful mutilations. They mirror the consequence of adverse events that may have dragged a person into depression. In the meanwhile, being impediments in completing tasks, they symbolize the fact of losing the will to act, a feature that often affects who is depressed. To conclude, the eye patch interferes with the ability of seeing correctly, being a translation of the distorted view of reality that often characterizes who struggles with depression, and tends to see everything negative and life events as insurmountable.

Story and Missions The story told and the missions that players are asked to complete are metaphors of the disease and are transpositions of its symptoms and features. For example, the mission in the Bermuda Triangle represents the vortex of depression and the difficulty of understanding what to do; the mission is based on messed information to interpret before providing the answer that unlocks the next step of the game. A further example is the encounter with the Sirens who sing to misdirect players and make them fail in answering the question on the smartphone; this mission represents the self-conviction of always being the unfortunate typical of whom is depressed.

Fig. 6.6 Four characters of *The Treasures of Captain Torment* with their representative objects (Boni et al. 2015)

Mechanics The game rules and the way players are allowed to conduct their activities were all designed as metaphors. The mechanic of the boat mirrors the importance of progressing together, without letting anyone behind, even if at a decisive moment this point will be challenged to put the players' morality to the test.

Places Since LBMGs are geolocated, the various PoIs throughout which the game unfold can be identified according to situated, specific meanings, in spite of being randomly picked because of their spatial position, or conveniently chosen because of their mutual distance. For example, the initial missions of the game are located in populated areas of the campus, while the final parts are in places progressively more isolated to symbolize a growing estrangement from the society, until the last mission that happens on a raised and remote ground, to represent the final separation from everything else.

Since 2013, to design LBMGs able to engagingly and effectively convey information about social issues and nurture reasoning accordingly, we employed practices now framed and known as obfuscating and distancing—as named by Kaufman et al. (2015). However, in parallel to such practice, we largely counted on the application of Goffman's concepts of *keying* and *fabrication* (1974).

Goffman describes *key* as a transcription or transposition, and hence the process of keying outlines the activity of transcribing or translating something to a new/ other level. As a consequence, Goffman's *keying* refers to the set of conventions that allows to frame a specific activity—already significant by itself according to a defined primary structure—into something *else*, acquiring a further meaning and appearing *translated or transposed* to those who look at it. This translation or transposition is grounded on laminations (layers) of meanings that lie on a primary framework of experience. Such structure can be accessed and understood just through framing it retrospectively. In consequence, *up-keying* is the process of introducing laminations, layers expressing and signalling a "variation of tone" in the discourse; while *down-keying* identifies the process of taking such layers away to reveal the primary structure and the *real* subject that has been framed. Examples of keying are make-believe, contests, ceremonials, or copies (Goffman 1974), namely situations where meanings go through processes of transformation that alter their appearance but maintain their original sense unaltered. In our case, the full set of metaphors employed assumes the function of keys; in parallel, the designer process that brings to the definition of such metaphors consists of an up-keying, while the player's practice of unpacking and extract meanings is a down-keying. A process that depends on players' ability to provide new sense to what they did, by quickly reviewing their entire game activities in the light of the new frame acquired.

To complete the rumination, we consider necessary to stress a broader reference to how the concept of key and games are connected. Keying is indeed functional to depict the behaviour of certain games (as persuasive games or games for social change) that push the act of representing forward, employing metaphors to make something appear as something else. This is considering that every game, as a simulation or interactive representation, is a reduction of the reality, and as such it can be considered a keying.

However, keying is not the only way to transform or re-frame a meaning. Goffman's concept of *fabrication* (1974) results indeed particularly fitting our needs of embedding meanings by obfuscating or taking the distance from their original, straightforward significance. Fabrication is the practice of inducing third persons (players in our case) to have a false belief about what is going on. Namely, what happens through those games that disguise the real topic addressed to eventually reveal it during the game, or once the game is over. In consequence of such disclosure, the meanings embedded in the game (in its various levels) can be unfolded and grasped. Accordingly, the acquisition of this renewed frame of meaning changes the perspective on and of the game itself, its elements, and even the experience is produced; this acquisition is such perspective-changing that prevents players from playing again as they did before. As said above, choosing to employ similar mechanics, including a final revelation that provides the game a new significance adducing further meanings, has clear impacts in terms of game replayability.

Taking this into consideration, and recalling the case study presented above, we can say that the totality of the game elements and features were designed to result highly metaphorical, employing in turn processes of keyings and fabrications. Accordingly, they consist of the designers' interpretation of specific significance that can be grasped and translated only in the light of an interpretation that unfolds further, revealing meanings. However, as said in Chap. 5, working out what the designers encoded and embedded in the different elements of the games is a process rather than straightforward.

It is common knowledge that stories are traditional means for expressing values, principles and significance, as well as for creating common views on specific topics or subjects (Davenport 2005). However, stories do not provide a general, *super-partes* perspective, but they always convey information from the point of view of whom is designing them. Analogously games that use stories to convey meanings transfer specific perspectives, the designer's ones. Acknowledging the complexity of decoding the message(s) embedded in the articulated system of meanings of the game, when a game is meant to convey specific pieces of information, perspectives or concepts, it is necessary to verify through several playtests first and game sessions later, that what the designer intends to convey actually corresponds to what is really grasped. As said explaining the methodology of our research, also this LBMG has been enquired conducting qualitative and quantitative research. The data culled via questionnaire and interviews that support that players were effectively able to decrypt the meanings and unpack the diverse levels of metaphors embedded in the game are summarized in the article *Empowering Games. Meaning Making by Designing and Playing Location Based Mobile Games* (Mariani and Spallazzo 2016), and in Mariani's Ph.D dissertation (2016). In synthesis, from questionnaire it resulted that the game largely benefitted of merging digital and non-digital elements, and in designing game tasks that required a real interaction between the two spheres. In this LBMG metaphors simultaneously lie on in-game physical and digital objects, game elements, as well on the overall story; moreover, always supported by the narrative, they are used to wisely bridge the digital and physical spheres, creating a continuity of meanings and sense between them. Then, data reinforced the very point on the ground of this chapter, namely that the fictional world chosen, the performance triggered, and the scenic material covers a crucial role in nurturing immersion and facilitating the exploration of the message. Particularly interviews revealed that once grasped the real topic addressed by the game, the simplicity and elegance of the idea on the ground of the game itself and its metaphors contributed to easily puzzle out the meanings behind the diverse elements.

References

Aarseth EJ (1997) Cybertext: perspectives on ergodic literature. Johns Hopkins University Press, Baltimore

Adams E (2010) Fundamentals of game design. New Riders, Berkeley

Bateson G (1956) The message "This is play". Group Process 2:145–241

Bateson G (1972) Steps to an ecology of mind: collected essays in anthropology, psychiatry, evolution, and epistemology. University of Chicago Press, Chicago

Bertolotti E, Daam H, Piredda F, Tassinari V (2016) The pearl diver: the designer as storyteller. Desis Newtork, Milan

Bogost I (2007) Persuasive games: the expressive power of videogames. Mit Press, Cambridge

Boni A, Frizzi G, Taccola S (2015) The treasures of captain torment [LBMG]. Politecnico di Milano, School of Design, Milan

Brathwaite B, Sharp J (2010) The mechanic is the message: a post mortem in progress 311–329

Caillois R (1958) Man, play and games. Free Press, Chicago

Chatman SB (1980) Story and discourse: narrative structure in fiction and film. Cornell University Press, Ithaca

Crawford C (2004) Chris Crawford on interactive storytelling. New Riders, Berkeley

Dansky R (2014) Screw narrative wrappers. In: Gamasutra. https://www.gamasutra.com/blogs/RichardDansky/20140623/219615/Screw_Narrative_Wrappers.php. Accessed 29 Dec 2017

Davenport G (2005) Desire versus destiny: the question of payoff in narrative. http://ic.media.mit.edu/people/gid/papers/DesireVsDestiny2005.pdf Accessed 30 Apr 2015

de Souza e Silva A (2006) Re-conceptualizing the mobile phone—from telephone to collective interfaces. Society 4:108–127. https://doi.org/10.1177/0094306106035000209

de Souza e Silva A (2017) Pokémon Go as an HRG: mobility, sociability, and surveillance in hybrid spaces 5:20–23

Dena C (2017) Finding a way techniques to avoid schema tension in narrative design. Trans Digital Games Res Assoc 3(1):27–61

Deskins TG (2013) Stereotypes in video games and how they perpetuate prejudice. McNair Scholars Res J 6(1):5

Frasca G (2001) Rethinking agency and immersion: video games as a means of consciousness-raising. Digital Creativity 12(3):167–174. https://doi.org/10.1076/digc.12.3.167.3225

Frasca G (2003) Simulation versus narrative. In: anonymous the video game theory reader. In: Wolf MJP, Perron B (eds) Routledge, New York, pp 221–235

Fornaro S, Lancini V, Spalenza A (2015) The divine tragedy [LBMG]. Politecnico di Milano, School of Design, Milan

Gee JP (2003) What video games have to teach us about learning and literacy. Computers in Entertainment 1(1):20

Goffman E (1974) Frame analysis: an essay on the organization of experience. Harvard University Press, Cambridge

Huizinga J (1938) Homo Ludens, 2002 edition. Giulio Einaudi Editore, Torino, Italy.

Joseph C (1949) The hero with a thousand faces. Princeton University Press, Princeton

Juul J (1999) A clash between game and narrative. Master thesis, University of Copenhagen

Juul J (2001) Games telling stories. Games Studies 1(1)

Juul J (2005) Half-real: video games between real rules and fictional worlds. MIT Press, Cambridge

Juul J (2008) The magic circle and the puzzle piece. In: Conference proceedings of the philosophy of computer games 2008. University Press, Potsdam, pp 56–67

Kaufman G, Flanagan M, Seidman M (2015) Creating stealth game interventions for attitude and behavior change: an "Embedded Design" model. In: Proceedings of the 2015 DiGRA conference 12, Article 102. http://www.digra.org/digital-library/publications/creating-stealth-game-interventions-for-attitude-and-behavior-change-an-embedded-design-model. Accessed 10 Jan 2017

Laurel B (1991) Computer as theatre: a dramatic theory of interactive experience. Addison-Wesley Longman, Boston

Lebowitz J, Klug C (2011) Interactive storytelling for video games: a player-centered approach to creating memorable characters and stories. Taylor & Francis, London

Mariani I (2016) Meaningful negative experiences within games for social change. Designing and analysing games as persuasive communication systems. Dissertation, Politecnico di Milano

Mariani I, Gandolfi E (2016) Negative experiences as learning trigger: a play experience empirical research on a game for social change case study. Int J Game-Based Learn 6(3):50–73

Mariani I, Spallazzo D (2016) Empowering games. Meaning making by designing and playing location based mobile games. ID&A Interact Des Archit 28:12–33

Mariani I, Spallazzo D (2017) Interactive players. LBMGs from a design perspective. In: CEUR workshop proceeding 1956:1-6. CEUR workshop proceedings

McMahan A (2003) Immersion, engagement and presence. In: Wolf MJP, Perron B (eds) The video game theory reader. Routledge, London, pp 67–86

Montola M, Stenros J, Waern A (2009) Pervasive games. Experiences on the boundary between life and play. Morgan Kaufmann Publishers, Burlington

Murray JH (1997) Hamlet on the Holodeck: the future of narrative in cyberspace. MIT Press, Cambridge

Myers DG (1987) Social psychology. McGraw-Hill, New York

Ryan M (2006) Avatars of story. U of Minnesota Press, Minneapolis

Ryan M (1991) Possible worlds, artificial intelligence, and narrative theory. Indiana University Press, Minneapolis

Ryan M, Thon J (2014) Storyworlds across media: toward a media-conscious narratology. U of Nebraska Press, Lincoln

Salen K, Zimmerman E (2004) The rules of play: games design fundamentals. MIT press, Cambridge

Sicart M (2011) Against procedurality. Game Stud 11

Spallazzo D, Mariani I (2017) LBMGs and boundary objects. Negotiations of meaning between real and unreal. In: Proceeding of the 6th STS Italia conference

Stenros J (2014) In defence of a magic circle: the social, mental and cultural boundaries of play. Trans Digit Games Res Assoc 1. https://doi.org/10.26503/todigra.v1i2.10

Venditti S (2017) Social media fiction. A framework for designing narrativity on social media. Dissertation, Politecnico di Milano

Chapter 7
Beyond the Digital: Reflecting on Objects and Contexts

Abstract The chapter addresses the role of physicality in LBMGs exploring how game objects may translate the fictional world into the real one, acting as boundary objects that intertwine the two worlds, rather than simply overlapping them, and activate negotiation of meaning. Furthermore, it analyses how the physical context may influence the game experience and how it may be included by designers in LBMGs.

Keywords Physical Game Elements · Boundary Objects · Bodily Engagement
Agency

Games convey meanings, because they are innately fraught with sense. Recalling Huizinga (1938), games and play are deeply intertwined with human society and the creation of meaning. Games may act as contexts of representation (Frasca 2003; Salen and Zimmerman 2004) wherein designers can embed meaning (Flanagan and Nissenbaum 2014) then translated by players via subjective interpretation (Sicart 2011). A meaning can be embedded in the interactions between players and the surrounding space, and with the game objects, which can act as boundary elements, intertwining the fictional and the physical world.

As a matter of fact, the rich interaction among players and with the surrounding space that very often characterizes LBMGs—and in particular urban event-based LBMGs—may be triggered by game objects. Here, we analyse physical game elements for their agency (Latour 2007) on players, that is delegated to them (Kaptelinin and Nardi 2009) by game designers. Agency can be interpreted as the ability of an agent to act, and therefore as the ability to produce effects in the world. Accordingly, every existing thing, being it an object, an animal or a person possesses agency. Looking at the interaction between players and objects as an exchange of meaning, the role of the designer emerges as that third pole, whose agency is embedded in the game and arises thanks to the game objects and the mechanics.

In game studies, the concept of agency has its roots in the video game field and has been investigated as the manner computer gamers play games as simulations

© The Author(s), under exclusive licence to Springer International Publishing AG, part of Springer Nature 2018
D. Spallazzo and I. Mariani, *Location-Based Mobile Games*,
PoliMI SpringerBriefs, https://doi.org/10.1007/978-3-319-75256-3_7

(Murray 1997; Sect. 6.1.1), studying the role of this medium in raising consciousness (Frasca 2001). Nevertheless, we expand the discourse to LBMGs, analysing how they manage to engage players with real objects, involving them into less-mediated processes of sense-making (Wright et al. 2008).

Hybrid Location-Based Mobile Games, indeed, lead players to roam the urban space exploring it, looking at things and attributing them a sense, in an unusual way. In such context, the ordinary physical elements may acquire a further meaning, contributing to creating novel consciousness, and act as tools players may use to empower themselves (Wertsch 1998).

The role of physical game elements as triggers of players' actions in LBMGs is still partially unexplored. In the chapter, we interpret them as boundary objects, able to link the fictional and the physical worlds, and to translate meanings across the two realms.

7.1   Bounding Two Worlds Through Physical Game Elements

LBMGs designers frequently overlook or neglect the role of physical game elements, mostly focusing on the hybridization between fictional and real world through the mobile device screen, by overlaying information, tasks and game mechanics on reality. A condition that is compulsory for *global* LBMGs (Montola 2011), since they must be played everywhere, but easily deployable in *local* and especially in event games.

The games analysed in this study were designed integrating physical game elements into the play experience, intertwining the praxis of urban games, usually enhanced with props and game objects, with that of LBMGs, mostly technology sustained (Montola et al. 2009). The goal for students was to fully exploit the features of both the worlds wherein players performs, namely the digital one, accessible through the mobile device, and the physical one.

Given these premises, physical game elements were employed both as means to spark players' interactions and as boundary objects, here discussed following Star and Griesemer's interpretation (1989), as flexible objects able to adjust to different worlds without giving up their common identity, or at least preserving coherence across the two realms.

The physical objects available to students for designing their games were objects already existing in the environment, ordinary objects placed by students into the playground and bespoke object, designed to carry out a specific task in the game.

Following the categorization of boundary objects proposed by the above cited authors (Star and Griesemer 1989), physical game elements in LBMGs can be included in all the four groups. Indeed, they can be urban objects borrowed from a stack (Star and Griesemer 1989) by designers and players to create their own urban *repository* (i) but they can also be arranged as (ii) *ideal boundary objects*, namely objects indefinite enough to be easily fitted in both the two worlds.

Physical game objects can be also considered (iii) *coincident boundaries*, objects that acquire a different meaning according to the world wherein they are employed. By keeping their meaning unaltered regardless of the context, they can also configure as (iv) *standardized forms*. This last category is also identified by Bruno Latour with the name of *immutable mobiles* (Latour 1986).

By bounding the two realms wherein the game takes places, physical game elements work as interfaces and, in doing so, they can influence our daily life, or play experience, modifying our perception of the surroundings in terms of space and social interaction (Johnson 1997).

Clearly, the same role may be played by digital devices. As conceptualized by de Souza e Silva (2006), arguing about devices as social interfaces and highlighting the role of the cultural context in defining interfaces, devices mediate relationships between users, and between users and their surroundings. As a matter of fact, the social meaning of interface is the result of how it is integrated into social practices and is not only reliant on technology itself.

7.2   Physical Objects as Triggers of Players' Interactions

The reasoning on the aesthetic experience between individuals and technology advanced by Wright et al. (2008), as well as the holistic approach that distinguishes an activity where every game element—physical or digital—is interrelated and understandable only when referenced to the whole, is the theoretical premises for our understanding of players' interaction with physical elements during the gameplay.

Their role in triggering interactions is due to the way they are interpreted by players in terms of meaning, since different readings may result in diverse strategies of action. Consequently, objects interpretation may impact on the overall game consistency and precisely on how the game bears its meaning. At the basis, there is the postulation that players make sense of game objects and negotiate their meaning relying on their personal and cultural background, an inclination further encouraged by the nature of these games that ask players to decipher, interpret and transfer their understanding from the in-game context to the real world.

The analysis on how game objects relate with the surrounding and with players moves from the assumption that players, objects and settings contribute to trigger a metaphorical dialogical or relational approach (Wright et al. 2008) to produce sense.

7.2.1   Game Elements and the Surroundings

Moving from the study of the sample of 44 LBMGs, we can identify three main categories of physical game elements, differentiated on how they relate with the space wherein they are set and how they are understood by players. A first category, *ordinary objects* (a), describes those objects that are well integrated in the surrounding and that acquire a different meaning for players only. The second typology, *common objects* (b), as the name suggests, is populated by ordinary objects that can be recognized as out of context by both players and non-players. The last category, *bespoke objects* (c), contains objects designed specifically for the game and, as such, very likely recognized as "outsiders".

The physical game elements pertaining to first category (a) do not declare themselves as part of the game, since they are already part of the gamespace or placed there purposely, but perfectly integrated. As an example, a trash bag on a bin may be interpreted by non-players as a forgotten object, but it can be useful and meaningful for players who temporarily live across the boundary of the real and fictional worlds. This is what happens in *The Fellowship of the Umbrella* (Bianchini et al. 2014), a LBMGs already discussed in Chap. 5, where the bin acts as a powerful tool, since it allows one of the characters, the Powerful Beech, to move by wrapping it around the legs. The sack-race-like movement makes players physically experience the fatigue of moving, unwittingly playing the role of a person with motorial disability (Fig. 7.1).

The sack, as well as similar objects, can be classified as *coincident boundary* in the categorization of Star and Griesemer (1989), since they acquire a diverse meaning according to the world wherein they are used and interpreted. Indeed, they do not have any impact or influence on non-players, but they become meaningful for players because of the action they enable in the gameplay and because of the meaning they embed.

Objects potentially recognized as out of context by both players and non-players define the second category, *common object* (b), exemplified in the game *Drop.it* (Bonfim et al. 2016): water filled glass ampoules filled hang from trees and must be collected by players (Fig. 7.2). While in a kitchen some glass ampoules would not

Fig. 7.1 A player personifying the *Powerful Beech* in *The Fellowship of the Umbrella* (Bianchini et al. 2014)

Fig. 7.2 Ampoules employed in *Drop.it* (Bonfim et al. 2016)

be noticed, on trees they can be immediately understood as out of context. Both non-player and players can recognize them as out of context, but only players can fully get their role and meaning in the gameplay. For outsiders, the ampoules could be alternatively a prank, a work of art, or an advertising stunt. In another game, the above cited *The Fellowship of the Umbrella* (Bianchini et al. 2014), players find a mask with snorkel in the university campus, where no beaches and sea can be enjoyed by students. Consequently, the mask, that is a tool for the Magician, can be evidently recognized as out of context. Objects pertaining to this category, here exemplified with the ampoules and the mask, on the one hand, are easily recognizable as weird, out of what people would expect in that space, but, on the other, do not necessarily declare that a game is going on. They actually work as boundary objects for players, linking the real and the digital realms and, in doing so, augmenting their immersion in the hybrid world of the game. However, they do not achieve the same result with non-players: they are not powerful enough to make them perceive the surrounding as a playground.

Physical game elements purposely designed and realized for the game, and recognizable as part of it, characterize the last category of physical game elements, *bespoke objects*. The game *The Origins of Forging* (Belloni et al. 2016) makes broad use of these objects, colonizing the playground with cardboards portraying wild animals that acquire a specific role in the gameplay. Such objects pop up from the context so fulsomely that they are clearly identified as outsiders both by players and non-players. Independently from their in-game role, these objects are understood in the same way, and their effect on the worlds is identical; moreover, they convey the same information regardless of the context, configuring themselves as *standardized form* (Star and Griesemer 1989) or *immutable mobiles* (Latour 1986). Nevertheless, a difference is kept, since players must interact with them in order to proceed with the game, while they act as traces for non-players, physical hints that let them understand of being in a space that is a hybrid playground, where something is happening.

Three different behaviours can be recognized if we analyse the above-mentioned three categories of physical game elements in respect to the concept of magic circle

(Huizinga 1938; Caillois 1958; Salen and Zimmerman 2004). *Ordinary objects* (a) keep players apart from non-players; in doing so, they do not blur the boundaries of the magic circle. Non-players are unaware that a game is going on. Those traits of separateness and secrecy outlined by Huizinga (1938) are preserved, and the urban space is perceived as a playground by players only. *Common objects* (b), instead, can contribute to weakening the borders of the circle. Being evidently perceived as out of context, they can let non-players understand that something out of the ordinary is going on, even if not clearly related to a game activity. *Bespoke objects* (c) clearly declare "this is play" (Bateson 1972) and expand spatially the borders of the magic circle (Montola et al. 2009). While assuming a role for players, these objects make the game pervasive and become evidences of the fictional world superimposed on the real one.

7.2.2 Game Elements and the Players

Physical objects in LBMGs do not only interact with surroundings; their core role, indeed, is that of being used and interpreted by players. Accordingly, this section focuses on the relationship between players and physical game elements.

Our empirical study outlined different typologies of interaction between players and objects, here synthesized in three main categories described in the following. Players can interact directly with objects that do not bring any metaphorical meaning: we call this kind of relationship *direct interaction* (a). The manipulation and use of other kinds of objects may instead have an impact on the game in the digital dimension, configuring as a *mediated interaction* (b). Lastly, we define *metaphorical interaction* (c) the relationship with objects that act as substitutes or metaphors of fictional elements.

Serving in the real world is the main aim of objects characterizing the first typology of interaction (a): they are employed as part of the gameplay and support players to proceed the game. These objects show their plain function and are used according to their original purpose: in *The Origin of Forgings* (Belloni et al. 2016), players must collect and then assemble objects; in *The Treasures of Captain Torment* (Boni et al. 2015), the cardboard boat is a game prop; and in *SOS-Rescue Squad* (Panza et al. 2016), players circumscribe spaces with tapes to keep outsiders far away (Fig. 7.3).

Physical game elements have an impact on the digital realm if the interaction is *mediated* (b): players' actions in the real world modify what happens in the digital one. Clear examples are objects containing codes to be typed into the digital interface of the game to unlock levels or quests. In the LBMG *The Rapture* (Conti et al. 2015), objects such as a piñata, a styrofoam car and balloons must be destroyed or torn down to get alphanumeric codes. Other times bespoke objects fabricated via rapid prototyping are used as triggers and employed as vessels of codes that unlock elements in the digital dimension once recomposed, as it happens in *SOS-Rescue Squad* (Panza et al. 2016).

Fig. 7.3 From the top left: the cardboard boat of *The Treasures of Captain Torment* (Boni et al. 2015), players assembling objects in *The Origin of Forgings* (Belloni et al. 2016) and the bench circumscribed with the tape in *SOS-Rescue Squad* (Panza et al. 2016)

In the same game, players must encircle a bench with tape and in doing so, they modify its usual, social function. The bench becomes of a space of non-sociality and acquires a metaphorical meaning (Fig. 7.3). The object becomes symbolic of something else with a role in the fictional world, configuring this interaction as *metaphorical* ©. Analogously, bananas and coffees become guns and poisons in several assassination games—recalling the pervasive game genre as discussed by Montola et al. (2009).

7.3   Attitudes of Employing Physical Game Elements

The discussion about LBMGs physical objects as influencers of behaviours and triggers of meaning-making in respect to the surroundings and players has been presented in the previous sections. At this point, we consider necessary to move the discussion on how the agency of these objects impacts on different aspects of the gameplay, activating (1) social engagement, (2) negotiation of meaning and (3) in-game behaviours.

As emerged in Chap. 4 discussing the social configuration of players, some physical objects may play a relevant role in fostering social engagement and interpersonal connections during the play experience. The already mentioned cardboard boat of *The Treasures of Captain Torment* (Boni et al. 2015), beyond its function of game prop as previously discussed, makes players squeeze in a limited space, and, in doing so, it enforces the sense of community while making the game manifest to passers-by. Moving across the campus in a cardboard boat is clearly an awkward behaviour, accepted only because recognized as part of a ludic activity.

The existence of the magic circle and the detachment of what is play from the ordinary life are reinforced by its recognition, and physical objects work as masks or identity items. They acquire a relevant role in communicating to other people that "this is play", and it is happening here and now (Bateson 1972). These masks serve a social role: who wears them is clearly recognized as belonging to a group or "community" (Caillois 1958). Objects become an overt statement of the presence of the magic circle, and may increase the possibility to achieve immersion (Murray 1997), increasing players' consciousness of being involved in the game world and its story (Mcmahan 2003).

Then, other game objects force the expansion of the magic circle (Montola et al. 2009) by favouring the interaction between players and non-players. The stickers of *The Infection* (Bassanese et al. 2015), discussed in Chap. 4, force social interaction and require players to go beyond their usual comfort zone, challenging the interpersonal distance, by asking them to place stickers or explore players' body.

Game elements, therefore, act as boundary objects and activators of a certain negotiation of meaning between real and "unreal", transferring and translating the fictional world into the real one. As a result, the two worlds are intertwined, rather than simply overlaid. Studying the sample of 44 LBMGs object of study, we recognized three different attitudes in using physical objects with the aim of embedding meaning.

The first attitude, that we call *transfer*, is based upon the use of physical objects to augment the immersion into the fictional world (Murray 1997). These objects, very often created by the designers, are the concretization of the fictional world into the physical one: objects thrown in the real world that act as boundaries between the two realms. Examples coming from the aforementioned LBMGs are the cardboard boat of *The Treasures of Captain Torment*, the spiked gloves of the *The Rapture* or Hermes' winged sandals in *The Origins of Forging*.

Translate is the second recognized attitude, and it proceeds by embedding metaphorical meaning into objects. By doing so, a double operation of translation is required: designers employ objects chosen or created for their symbolic meaning, and players should understand it to proceed with the game or make sense out of the experience Sect. 5.3) (s). The meaning embedded by designers into the objects with a first operation of translation is then decodified by players during the play experience (see Sect. 5.2). Referring to the studied LBMGs, a bag trash becomes a metaphorical representation of handicaps, by setting players in the shoes of people with motor disabilities, while a mask with snorkel represents the condition of being dumb (see Sect. 7.2.1).

Players must negotiate the meaning of these objects in order to fully get not only their functional role in the gameplay but also their hidden message. And this progressive negotiation could be a step-by-step revelation of the real meaning of the objects, or of the game itself, or it could be an epiphany at the end of the game, following the stealth approach proposed by Kaufman and Flanagan (2015). The revelation could also activate a process of translation and reinterpretation of the experience *ex post*.

The third attitude matches the previous two and accordingly is named *transfer and translate*. This approach describes the use of physical game elements that both transfer the fictional world into the real one and embed meaning. An example is the use of the smartphone that can be in the meantime a way of mixing the real world with the fictional one and a metaphorical object, becoming a magic tool in some games, and even embodiment of characters such as narrators and guides, in others.

The analysed LBMGs offer examples in this sense. In the *The Rapture* (Conti et al. 2015), a styrofoam car must be destroyed by players to get alphanumerical codes. The car is clearly a physical representation of the fictional world, but the act of wrecking is also symbolic of the violence players are unconsciously committing (Fig. 7.4). Hence, physical game objects can be together triggers of immersion in the fictional world but also conveyers of meaning, and in so doing, they can be considered as powerful tools in the kit of LBMGs designers.

The last point to be discussed regards the role of physical game elements in triggering in-game behaviours, persuading (Redström 2006) players to perform the expected actions. In crafting artefacts that will serve the game, designers have expectations for the objects capability to prompt the desired gameplay. On player's side, their actual experience must be examined to confront how such objects have been employed and understood in respect to designers' expectations. Clearly, players can both confirm or contradict the initial expectancies, and designers have the specific duty to understand if players comprehended correctly the function and the meaning of each object.

Our three-year-long study on LBMG that incorporated physical game elements situated in the space shows that physical game elements can be for designers both allies or enemies, on the basis of how they are designed and provided with meaning, as well as to how they are interpreted.

Physical game elements, indeed, if wisely designed can be allied to designers in fostering the desired gameplay and embedding meaning. By acting as boundary objects, they play a double function of linking the fictional world with the real one, and becoming concretizations of the meanings designers aim to transfer. On the contrary, if misused, they can damage the game by suggesting an inconsistent gameplay or by miscommunicating its contents and meanings. Consequently, designing LBMGs disregarding the potential of such objects as triggers of

Fig. 7.4 Physical game elements to be destroyed in *The Rapture* (Conti et al. 2015)

behaviours and communication means can jeopardize or damage the play experience, configuring objects as enemies. As demonstrated, physical game elements can be influential in modifying their relationship with players and consequently affecting the way their meaning is interpreted.

The categories discussed in this chapter are intended to serve as a framework to empower designers involved in the creation of LBMG that include physical game elements. Clearly, our discourse is limited to local LBMGs (Montola 2011) and in particular to event games that can count on a rich set of props and bespoke objects.

The ambiguity of physical game elements, that can serve both as allies and enemies, is a matter of design, and must be coped foreseeing players' in-game behaviours and their interpretation of the embedded message.

References

Bassanese G, Bonfarnuzzo L, Pham C, Redana F (2015) The Infection [LBMG]. Politecnico di Milano, School of Design, Milan, Italy

Bateson G (1972) Steps to an ecology of mind: collected essays in anthropology, psychiatry, evolution, and epistemology. University of Chicago Press, Chicago, IL

Belloni E, Bucalossi C, Mazzoleni C, Menini M (2016) The origins of forging [LBMG]. Politecnico di Milano, School of Design, Milan, Italy

Bianchini S, Mor L, Princigalli V, Sciannamè M (2014) The fellowship of the umbrella [LBMG]. Politecnico di Milano, School of Design, Milan, Italy

Bonfim Bandeira F, Marcon S, Namias C, Paris M (2016) Drop.it [LBMG]. Politecnico di Milano, School of Design, Milan, Italy

Boni A, Frizzi G, Taccola S (2015) The treasures of captain torment [LBMG]. Politecnico di Milano, School of Design, Milan, Italy

Caillois R (1958) Man, play and games. Free Press, Chicago, IL

Conti N, Saracino G, Serbanescu A, Valente N (2015) The rapture [LBMG]. Politecnico di Milano, School of Design, Milan, Italy

De Souza e Silva A (2006) Re-conceptualizing the mobile phone—from telephone to collective interfaces. Society 4:108–127. https://doi.org/10.1177/009430610603500209

Flanagan M, Nissenbaum H (2014) Values at play in digital games, Reprint edn. The MIT Press, Cambridge, MA

Frasca G (2001) Rethinking agency and immersion: video games as a means of consciousness-raising. Digit Creativity 12(3):167–174. https://doi.org/10.1076/digc.12.3.167.3225

Frasca G (2003) Simulation versus narrative: introduction to ludology. In: Wolf MJ, Perron B (eds) The video game theory reader. Routledge, New York, NY/London, UK

Huizinga J (1938) Homo Ludens, 2002nd edn. Giulio Einaudi Editore, Torino, Italy

Johnson S (1997) Interface culture: how new technology transforms the way we create and communicate, 1st edn. Harper San Francisco

Kaptelinin V, Nardi BA (2009) Acting with technology: activity theory and interaction design. The MIT Press, Cambridge, MA/London, UK

Kaufman G, Flanagan M (2015) A psychologically "embedded" approach to designing games for prosocial causes. Cyberpsychol: J Psychosoc Res Cyberspace 9

Latour B (1986) Visualization and cognition: drawing things together. In: Kuklick H (ed) Knowledge and society studies in the sociology of culture past and present. Jai Press, Greenwich Conn

Latour B (2007) Reassembling the social: an introduction to actor-network-theory, 1st edn. Oxford University Press, Oxford, NY

Mcmahan A (2003) Immersion, engagement, and presence: a method for analyzing 3D videogames. In: Wolf M, Perron B (eds) The video game theory reader. Routledge, pp 67–86

Montola M (2011) A ludological view on the pervasive mixed-reality game research paradigm. Pers Ubiquitous Comput 15:3–12. https://doi.org/10.1007/s00779-010-0307-7

Montola M, Stenros J, Waern A (2009) Pervasive games. Expreriences on the boundary between life and play. Morgan Kaufmann Publishers, Burlington, MA

Murray JH (1997) Hamlet on the holodeck: the future of narrative in cyberspace. MIT Press, Cambridge, MA

Panza M, Pozzi L, Rota P, Veschi D (2016) SOS-rescue squad [LBMG]. Politecnico di Milano, School of Design, Milan, Italy

Redström J (2006) Persuasive design: fringes and foundations. In: IJsselsteijn WA, de Kort YAW, Midden C et al (eds) Persuasive technology. Springer, Berlin Heidelberg, pp 112–122

Salen K, Zimmerman E (2004) The rules of play: games design fundamentals. MIT press, Cambridge, MA

Sicart M (2011) Against procedurality. Game Studies 11

Star SL, Griesemer JR (1989) Institutional ecology, 'translations' and boundary objects: amateurs and professionals in Berkeley's museum of vertebrate zoology, 1907–39. Soc Stud Sci 19:387–420. https://doi.org/10.1177/030631289019003001

Wertsch JV (1998) Mind as action. Oxford University Press, Oxford, NY

Wright P, Wallace J, McCarthy J (2008) Aesthetics and experience-centered design. ACM Trans Computer-Human Interaction 15:18:1–18:21. https://doi.org/10.1145/1460355.1460360

Chapter 8
Conclusions

Abstract It presents a critical synthesis of the arguments discussed throughout the chapters. It aims at focalizing the readers' attention on the major contributions of the book, advancing inflection for researchers, designers, practitioners and entrepreneurs in the field.

Keywords Pervasive Games · Immersion · Interaction · Meaning Making
Context · Situated Experience

This book reflects years of research and exploration. This could seem a straightforward reflection on game design as a practice that requires to puzzle multiple perspective to frame LBMGs that are as entertaining as meaningful to play, but it goes far beyond.

When we started to structure this book, we intended to collect the experience and expertise we gleaned from a rich field experimentation that counted on a significant series of games (44 LBMGs) designed over three academic years, and grounded on a rich and lively theoretical literature. However, the practice-based reflective insights gathered throughout these pages epitomize a research we started in 2012 to probe the potentialities and constraints of a field that is continuously fed by technological progresses.

We decided to start framing LBMGs from a design perspective, investigating their compelling contemporary practices and applications, and in so doing we soon failed to respect the borders of our own disciplines—communication and interaction design, game studies and game design and media studies; borders that are blurred by their own nature, and that once more stress how such games require to be investigated and designed adopting interdisciplinary frames of mind and practice.

What emerges is that bridging the real and the digital into a space that is hybrid and connected opens a multiplicity of design possibilities that goes far beyond new technological challenges, portraying LBMGs as a dense scenario in need of exploratory processes aware of their multifarious implications. In consequence, apart a first chapter that serves to contextualize our ruminations, we articulated our reasoning through six chapters devoted to exploring how LBMGs relate to as much

D. Spallazzo and I. Mariani, *Location-Based Mobile Games*,
PoliMI SpringerBriefs, https://doi.org/10.1007/978-3-319-75256-3_8

thick dimensions: from the new times and spaces of play (Chap. 2), to learning by designing and playing such challenging games (Chap. 3); from the uncommon social (Chap. 4) and communication (Chap. 5) dimensions that characterize LBMGs, to the narrative (Chap. 6) and sociological (Chap. 7) qualities they entail.

Each chapter provides interdisciplinary perspectives that stem from specific theoretical assumptions that we empirically applied to design case studies, tested conducting qualitative and quantitative research and collecting practical outcomes supported by data. It is the result of reasoning and experimentations we conducted over the years, relying on numerous investigations and iterations, implementation and verification, benefiting from interdisciplinary conversations, intellectual exchanges, as well as discussions among peers, academics, designers and researchers.

As a result, the multiple perspectives that constitute this book can be seen as sources of reflections for those researchers and scholars who investigate the field in its variety and complexity, as bricks of a curriculum that introduce LBMGs design practices to designers and practitioners, and as a repository of knowledge for entrepreneurs who want to broaden their scholarship about the real implications for LBMGs as communication systems.

One of the main points we gained while probing the contemporary LBMGs gaming practices is the role of mechanics in supporting engagement and facilitating immersion. Examining the games on the market, the field returned a panorama with a set of mechanics and goals that can be described as unsurprising, conventional in its variety, and often poor in originality. Hence, we focused on researching the elegance of the gameplay and the meaningfulness of the experience, looking for significant and pleasant interactions with the game as a complex system with mechanics, narrative and technology consistently and coherently intertwined. In particular, what emerged from our practice-based investigation is that LBMGs largely benefit from well-structured and fascinating narratives that evolve through the game and challenge players to enter unpredictable stories. Stories that involved players into challenging and engaging experiences that then revealed to be fraught with meanings. The LBMGs designed revealed indeed to be able to convey values of social matter, and put meanings at play succeeding in triggering new ways of understanding existing and contemporary issues—usually employing catchy and fashion narratives that use a stealth approach in order to distance the player from the real messages embedded, enhancing the communicative potentialities of the game experience.

In these stories, the surroundings, namely physical spaces where the game takes place, are more than mere environments throughout which players can surf: rather than being instrumental tools or mere playgrounds, they assume the role of in-game "actors", actively involved in communicating contents. This decision of empowering with specific meanings the places in which the steps of the games are set may appear obvious, but is actually rarely applied because of its design implications especially in terms of narrative consistency and game replayability.

A further element that we consider necessary to further stress, especially while wrapping up and discussing the constraints and potentialities of LBMGs, is the diverse way in which such games have an impact. In particular, we intend to unpack this point through a threefold perspective, considering their effect on the surroundings, on players and on non-players. LBMGs are pervasive games that take place in the urban space; some are designed to be played for short periods, others for longer spans of time; some come be played by single players, others require teams; and their purposes range from recreation to educations. As repeatedly stated, in our research and in this text we mainly focused on a side branch, since we focused on games that share several traits with events, that cover issues of social or cultural concern and that employ physical elements as game objects.

That said, starting from analysing how LBMGs impact on the gamespace where they are played, it immediately pops up that the game per se is quite stealth: it stays hid and quite until it is played, because in that moment, it spreads in the surrounding, becoming pervasive in a spatial, temporal and often social way. Especially, when it capitalizes on the cultural capital of play in public spaces and embed performative aspects, LBMGs can become triggers of complex social interactions that fully exploit both in-person social contacts—among players, between players and non-players—and the context of play, with its features and traits. It becomes clear that a game is happening, and that it is pervading the surrounding space. In doing so, it enacts a sort of re-appropriation of the public spaces by hands of players. Discussing pervasivity, we cannot help underlining that the expansions these games activate are often as unexpected as "disrespectful", since they make the game alive and present, whether those who are non-players—namely inhabitants or passers-by that are in the gamespace where it occurs—like it or not.

To optimize the game experience, we draw particular attention on those game objects that act as boundary objects, and intertwine the fictional and the real world, as well as the fictional and physical dimension of play. They are physical objects that solve particular narrative and in-game functions, and that contribute to increasing the game meaningfulness: they become evidences of the overlaps and expansions of games in the everyday life. From a design perspective, including such elements means designing LBMGs as urban events; as such they often require specific material to be staged, as well as designing particular dynamics of play. However, what is often underestimated is the social meaning of these games, namely the social configurations that such activities entail, the experience they generate, and how they can impact on both players and non-players. As a matter of fact, as a layer overlapped to the everyday life, these games temporarily alter our cities and the meanings we provide to their places/spaces.

The last rumination we want to address refers to the meanings these LBMGs can embed and their implications. From the players' perspective, these games are often "deceitful", since they speak of something meaning something else; hence, they become demanding towards players who are asked to make sense of the game by interpreting and decoding its layers of meaning—meanings that are simultaneously placed in the real as in the fictional worlds, and spread among the different elements of the game. In so doing, LBMGs players get challenged on a higher level, grounded on practices of interpretation and re-interpretation. Fully harnessing their technological potentialities, LBMGs turn into meaningful, dense field of experimentation where designers can codify, represent and perform meanings that players are asked to grasp and interpret. As might be expected, accepting as design drivers the inclusion of multiple layers of meaning, spread among the physical and digital realms, implies some relevant consequences: the phases of design and prototype, test and analysis, following an iterative process, have to be applied not only to the game "playability", but also to its ability to transfer the expected meanings.

Taking meanings into consideration implies adding design constraints and includes the player individuality and subjectivity as crucial variables in the process of conveying specific messages by embedding them into games.

Printed in the United States
By Bookmasters